LIGHTNING SKY

LIGHTNING SKY

A U.S. Fighter Pilot Captured During WWII and His Father's Quest to Find Him

R. C. GEORGE

Foreword by Marcus Brotherton

CITADEL PRESS
Kensington Publishing Corp.
www.kensingtonbooks.com

CITADEL PRESS BOOKS are published by

Kensington Publishing Corp.
119 West 40th Street
New York, NY 10018

ISBN-13: 978-0-8065-3896-9
ISBN-10: 0-8065-3896-1

First Citadel hardcover printing: May 2019

10 9 8 7 6 5 4 3 2 1

Printed in the United States of America

Library of Congress CIP data is available.

Electronic edition:

ISBN-13: 978-0-8065-3897-6 (e-book)
ISBN-10: 0-8065-3897-X (e-book)

For those who never came home

Oh, I have slipped the surly bonds of earth,
And danced the skies on laughter-silvered wings;
Sunward I've climbed and joined the tumbling mirth
Of sun-split clouds—and done a hundred things
You have not dreamed of—wheeled and soared and swung
High in the sunlit silence. Hov'ring there,
I've chased the shouting wind along and flung
My eager craft through footless halls of air.
Up, up the long delirious, burning blue
I've topped the wind-swept heights with easy grace
Where never lark, or even eagle, flew;
And, while with silent, lifting mind I've trod
The high untrespassed sanctity of space,
Put out my hand, and touched the face of God.

—JOHN GILLESPIE MAGEE, JR.
19-year-old American
with No. 412 Squadron,
Royal Canadian Air Force,
killed in action
on December 11, 1941

CONTENTS

FOREWORD

Every once in a while a new author emerges out of the blue and takes a genre by storm. With *Lightning Sky*, R. C. George has done exactly that.

For years I've worked closely with World War II veterans—the elite paratroopers and Marines featured in HBO's *Band of Brothers* and *The Pacific*. I'm particularly interested in stories that have never before seen the light of day, and I was excited to read this untold true story of a father's quest to rescue his Nazi-captured son—a twenty-year-old fighter pilot named Dave MacArthur, who was shot down over Greece in 1944.

The author's research is meticulous and will satisfy any ironsided military history reader, yet George is also a skilled chronicler who understands the delicate art of blending history with story craft. This nonfiction war narrative has the satisfying feel of a page-turning novel, and its aviation action sequences are among the best in the book world. George captures the voice of the characters and the flavor of the era in an authentic manner. Rich with metaphor and brimming with emotion, the pages are chock-full of patriotism, humor, and the resilient optimism that defined America's Greatest Generation.

Against the backdrop of war, when humanity is at its harshest, *Lightning Sky* offers a glowing reminder that some things in life are worth fighting for.

I hope you enjoy the book immensely.

> Marcus Brotherton
> *New York Times* bestselling author of
> *We Who Are Alive & Remain,*
> *A Company of Heroes*, and *Shifty's War*

A NOTE TO READERS

The story you are about to read is true.

As a steward of Dave MacArthur's story, and as a chronicler of history, I have attempted to transmit this narrative as faithfully as possible. No dialogue is created. Behind every comment in quotation marks, or thought in italics, is a historical document. The people, situations, and conversations come directly from interviews (most decades old), journals, letters, photographs, books, articles, and other reliable records, governmental and otherwise.

However, as in all retellings of nonfictional stories, the author is tasked with carving a path through the narrative.

Thus,

This story is not a complete one. . . . In war too many things happen too quickly to be recorded. All we can hope for is to recapture the highlights to help us recall those exciting days, places, incidents and people, and to live again in memory that period which, for most of us, was a high point in our lives.

—Major General John M. Devine, commander
of the Eighth Armored Division during World War II,
in his foreword to the 1956 book, *In Tornado's Wake*

R. C. George
January 2019

PART I

*For a pilot, every plane has its own personality,
which always reflects that of its designers
and colours the mentality of those who take
it into action.*
—PIERRE CLOSTERMANN,
French Spitfire pilot,
Royal Air Force

CHAPTER 1
ONE-MAN WAR

Five years after the end of WWII
The night of April 21, 1951, near Hwach'on Reservoir,
Fifty miles northeast of Seoul, South Korea

For the first time in his military career, Dave slept with his boots tightly laced.

It was approaching 11:00 P.M., and throngs of Chinese soldiers stirred in the hills. The full moon painted the Korean landscape with a ghostly glow—ideal conditions for a surprise enemy attack.

The small village of Wontong-ni veined between two towering hills. Dave would be an easy target for snipers. Rumor was the Communists were gearing up for a full-scale counteroffensive in their quest to besiege Seoul.

Ten months earlier, on June 25, 1950, the North Koreans had invaded their southern neighbors, instigating a war that would endure for three years. The United States and their United Nations allies entered the conflict to push back the enemy and rescue the non-Communist Far East outpost.

Most assumed the war would resolve quickly. But as the Cold War grew hotter, and the winter of 1951 thawed into spring, that dream dematerialized. President Truman had promised his troops that they'd be home by Christmas, but more than four months later the conflict still raged.

If the twentieth century had known a Third World War, the conflict would have come closest to igniting in April 1951 when China, backed by the nuclear-capable Soviet Union, launched the single greatest Communist assault in the Korean War. Three years earlier, the Soviets had successfully detonated their first atomic bomb, code-named "First Lightning." As a quarter of a million Chinese soldiers spilled into the Korean peninsula, which jutted out of China like an appendix ready to rupture, and as Soviet-built MiGs threatened U.N. air superiority, talk filtered positively up the U.S. chain of command about dropping thirty to fifty atomic warheads already being stockpiled at the Kadena Air Base in Okinawa. A *yes* from President Truman, who had already threatened their use, would launch the War Emergency Plan. Then, dozens of atom-splitters like Little Boy and Fat Man would transform the neck of Manchuria into a horizon of mushroom clouds.

Standing in the Communists' way was a long line of allied forces spackled thinly across the 190-mile-wide Korean Peninsula. The goal now was to drive the enemy back above the 38th Parallel.

Years later, the Korean conflict would be called the "Forgotten War," but to those fighting for a free-world democracy, the stench of corpses and the loss of brothers in arms would linger forever. By the end of April, a beleaguered but determined Western Force inched its fingers through the rugged ridges.

First Lieutenant David W. MacArthur was on the front lines of this coalition.

Standing six feet tall, and weighing 160 pounds, Dave

sheathed a mop of flaming-red hair inside his helmet. He had ice-blue eyes and a mouth quicker than a trigger. The ruddy New Englander had earned his pilot's wings as a teenager seven years earlier in Texas. He was now assigned to a position no pilot wanted: to serve on the ground as a Forward Air Controller with the Fifth Republic of Korea (ROK) Regiment, Seventh ROK Division.

After flying thirty-eight close air support missions with the Eighteenth Fighter-Bomber Group, the twenty-six-year-old relinquished his F-51 Mustang to march afoot with the 6164th Tactical Control Squadron, a ground-to-air communication branch of the 502nd Tactical Control Group. His equipment amounted to little more than a jeep, an interpreter, and a few enlisted men who knew how to operate the complicated cache of radio equipment.

The transfer was "the most dangerous job of the Korean War," according to one report, and could steal the wind from any pilot's wings. To be attached to an American division was perilous enough, but when tagged to a South Korean division, as Dave was, the appointment was tantamount to a suicide mission.

The United States established the Korean Military Advisory Group (KMAG) to offer allied leadership to the battle-green Koreans. In April 1951, 923 military advisors were sprinkled among the Korean ranks. Dave soon realized a better acronym, one unofficially coined by U.S. troops: Kiss My Ass Good-bye.

Most of the Koreans had worked as agrarians before the war, and didn't speak a word of English. Interpreters accompanied the KMAGs, but the Korean language itself lacked military jargon. Words like "machine gun" or "headlight," which had no Korean counterparts, had to be painstakingly described by interpreters as "the gun-that-shoots-very-fast" or "the candle-in-the-shiny-bowl." When interpreters were killed in combat, as they often were, KMAGs found themselves facing the enemy alone while the frightened ROKs fled for their lives.

Unable to fly, Forward Air Controllers faced a daunting task: to spot enemy movements on the front lines and help direct T-6 Mosquito air strikes against their targets. Dave had no infantry experience. Even the .45 pistol on his hip had been issued by the Air Force, not by the Army. It was the sort of mission that turned mild-mannered men into tigers and trigger-happy heroes into cowards. The risk of capture was high and the risk of friendly fire from the sky even higher.

Overnight, Dave was summoned out of the clouds. Even though most of his wingmen had been shot down, including his best friend, whom he had recently buried, trudging through unfamiliar mud with infantries would be no picnic. Flyboys thrived on speed, and pilots are never at home in the hills. At this stage of the war, however, the Fifth Air Force needed a few experienced aviators to delay their rotation schedule and even jeopardize their Air Force careers entirely. Without logging combat hours in the air, promotion wasn't possible. Nevertheless, Dave did his damnedest to make the best out of a bad situation. He accepted the transfer, trusting that nothing would be more deadly than an airman unafraid to put his boots on the ground.

One hour before midnight, a burst of gunfire jolted Dave from his sleep. He tumbled out of his sack and planted his boots on the floor. They were already laced, which had bought him a few precious lifesaving seconds. He snatched his belt, cocked his .45, and rushed out into the moonlight.

On any other night, he might have relished the pleasant ambiance. But as he filled his lungs with warm Korean air, his ears were bruised by the sound of bugles, whistles, drums, and shrieks. It was the sound of the enemy, and they were advancing. The cacophony spilled into the village of Wontong-ni. It was a nightmarish noise—a war technique trademarked by Chinese soldiers to psychologically disarm the enemy before an attack.

It worked. The Fifth ROK Regiment began to disintegrate as hordes of Chinese rushed madly in their direction. The allies were under heavy attack. And there were so many of them, hundreds, maybe even thousands—a sea of helmets flowing over the ridge.

So far in his stint as a Forward Air Controller, Dave had spent the first few weeks of April wiping out isolated Chinese units on a northern march. After a brief respite at Hyon-ri, he pressed north of Line Kansas, which lay beyond the 38th Parallel, southeast of the Iron Triangle. To Dave's southwest was the Hwach'on Reservoir. The villages surrounding the large body of water and its 256-foot-tall dam had been taken by allied forces, but some doubted that the thinly spread Sixth ROK Division could ward off an attack.

On the day before, April 20, he had noticed increased enemy activity in the northern hills along the Soyang River, when the sound of occasional rifle fire was eclipsed by the thunder of heavy artillery. Even so, Dave had coordinated eleven successful air strikes. Under his guidance, some fifty planes deposited their bombs and missiles in concentrated areas of opposition, shocking and demoralizing the enemy.

As day turned to dusk, Dave contacted a nearby relay plane to request air reconnaissance of the area. The news was sobering. Three Chinese divisions were moving toward his position, and they were carrying a mountain of artillery. How many soldiers? The pilot could only guess.

On the night of April 21, as the human tsunami crashed over the area, Dave found himself in the path of the greatest enemy assault in the Korean War. Approximately seven hundred thousand Communist soldiers, spread out among seventy divisions, rushed south to reclaim the capital.

The first thrust of the Chinese Fifth Phase Offensive had begun.

For the allies, the silver lining came always from the clouds. First came the whir of the wings, then the snarl of the engine, and suddenly an outfit of Reds discovered the horror of one thousand degrees Fahrenheit, nearly five times the boiling point of water. When dropped into valleys or spilled over ridges, the dreaded napalm, a jelly gasoline cocktail invented in a secret Harvard laboratory in 1942, could lick the bones of a hundred bodies and leave a cemetery of corpses frozen in white-hot hell.

The Chinese Red Army, though vastly more numerous, could not initially square up with the Far East Air Force, which sired a stable of jet-powered thoroughbreds. The Lockheed P-80 Shooting Star with its trailing mane of thin, black smoke. The mach-capable Republic F-84 Thunderjet. The elegant North American F-86 Sabre. Then there was the Boeing B-29, the mother of all bombers. When pregnant with a 12,000-pound Tarzon bomb, the Superfortress could birth catastrophe on bridges, supply dumps, and hydroelectricity facilities. The veteran B-29 had ended a war once, and given the green light, she could do it again.

In the meantime, the bulk of the attack against the Chinese came from an unlikely aircraft—the North American T-6 Mosquito. She was a single-engine trainer who, like Dave, acquired her wings in Texas. Neither of them, the man nor the machine, had any business leading a frontal assault in Korea.

With a maximum speed of only 210 miles per hour, the Mosquito was much slower than her fuel-thirsty colleagues. She lacked the innovative fuselages of the dual-piloted F-82

Twin Mustang. Her wings couldn't fold conveniently like those of the Vought F4U Corsair. Inverted flying was better suited for midwing fighters. If the Mosquito remained upside down too long, her belly would naturally topple back over like a toddler with a head too big for its body.

The Mosquito descended from a long line of training aircraft, beginning in 1935 with the NA-16 prototype. In World War II, she had graduated to the status of an "advanced" trainer, but most pilots moved quickly through her to mount loftier cockpits. To face off with a Soviet supersonic would spell trouble for the Mosquito. To take on an enemy MiG was a joke.

MiG Alley, that great sanctuary in the sky between the Yalu River and the Yellow Sea, belonged to better, faster warbirds with swept-back wings and thousands of pounds of thrust. With her meager six-hundred-horsepower engine, the "seeing-eye dog" struggled to find her howl beneath the jetted stratosphere.

But she did possess one advantage—the Mosquito could take a punch.

Behind her stubborn chin was a hardy fuselage capable of enduring substantial punishment from flak-heavy ground forces. Whereas jets were blinded by their speed, the Mosquito cruised at a comfortable 145 miles per hour, slow enough and low enough to look the enemy in the eyes. She could loiter at her leisure, turn on a dime, and had no problem threading a loose weave through the corrugated Korean hills.

Having trained in the Mosquito at Eagle Pass, Texas, Dave knew the plane's potential to summon death. Five months into the conflict, the Mosquito had earned the respect of every advancing U.N. troop. By April, the unlikely hero scored 90 percent of air-to-ground attacks and became the darling of Forward Air Controllers. By the end of the war, she would fly some forty thousand sorties. To the allies, her wings belonged to angels. To the Chinese, though, they buttressed fire-breathing dragons.

"Hold your positions!" Dave shouted at the panicked soldiers, but it was no use. The orchestra of bugles had reached a violent crescendo. The heavy artillery of the Fifth ROK Regiment soon answered back, but the enemy was close. The rigid slopes echoed with the sound of rifles and machine guns. The regiment was in disarray.

Some of the ROK soldiers scrambled for cover. Others instinctually threw up their hands to surrender as barrages of bullets peppered their bodies. Dave made a desperate beeline for his jeep. If he could reach his radio, a Mosquito might come to the rescue. He dodged rounds of incoming mortar, pausing only to return enough fire for his squad to secure the vehicle.

The entire area was now under assault. The Chinese reached the hut that Dave had abandoned only minutes earlier, and in the distance he could hear screams silenced by gunfire. To the east, a column of Chinese soldiers raced through the hills and the river valley to encircle the allies.

Dave assessed the situation. If the enemy gained a 360-degree advantage, they would choke off the only escape route leading to the southern city of Inje. Not even a low-flying strafe could offer much assistance if the Reds slipped their noose around the village and then tightened it. Five months earlier, the Chinese had employed a similar tactic at the Chosin Reservoir, when 120,000 Communists encircled 30,000 U.N. troops. The resulting seventeen-day battle claimed 10,495 lives. Dave couldn't let that happen. In the darkness he reached the jeep, cranked up the radio transmitter, and listened.

Static.

He tried again but was met with more static. Higher headquarters must be informed of the attack so they could dispatch immediate reinforcements, but the radio continued to bleed only the sound of static.

After two hours of unanswered calls, and with incoming fire drawing closer, Dave was running out of options. To save these men, he would need to somehow marshal his own air support. Overhead, a friendly Douglas C-47 Skytrain, affectionately called the "Gooneybird," ignited the night sky with bursts of orange.

"What the hell are you dropping flares on us for?" Dave screamed into the transmitter.

"You?" a pilot crackled back. "Hell, you're ten miles north of the lines."

Dave froze. Ten miles? Piece by piece, the battle scene materialized in his mind. To Dave's immediate west, the U.S. First Marine Division stood their ground, dishing out artillery as fast as the Chinese could absorb it. But the Sixth ROK Division had bolted to the rear of the fight. Their retreat created a two-mile gap in the allied line, a hole through which the Chinese quickly spilled.

If true, this was absolutely the worst-case scenario. "Tell the Joint Operations Center we have been cut off," Dave said. "Set up an airdrop of ammo and send in help as soon as possible."

A mob of confused ROK soldiers rushed to Dave's jeep. The horrified look on their faces suggested they were verging on collapse. They needed a leader, someone to guide them into battle or into exodus. For some reason, and despite his rank and lack of infantry experience, they identified Dave as that leader.

Fight or flee? It was a difficult decision to make, but the beleaguered men demanded an answer. Surprised by his own courage, Dave decided to stay and fight the enemy, at least until his ammunition ran out.

In that instant, and against the backdrop of an orange-striped sky, a man turned into a tiger. According to the official Air Force report, the F-51 pilot "suddenly became an infantry officer."

Seventy miles to Dave's west, the famed British Gloucestershire Regiment faced a similar situation. They were protecting a twelve-mile front north of Seoul when the Chinese descended on the regiment "like a swollen wave," as one young Gloster remembered. Unwilling to abandon their weapons, the British defended Hill 235 for three days, but when the ammunition ran dry the soldiers were surrounded and eventually overtaken. History would immortalize the "Glorious Glosters" for their heroic sacrifice at the three-day Battle of the Imjin River, when an army of 650 men staved off a legion of 10,000 Chinese soldiers.

Dave was made of the same stuff. He jumped into action, corralled the ROK soldiers into organized units, and then encouraged a handful of American military advisors to command the Koreans back to their posts. The artillery guns had to be remanned. The Chinese onslaught had to be checked.

A spray of small arms returned fire. Dave deployed more of the Korean troops, assigning them to strategic vantage points and ordering them to hold their ground at all cost. Before long, the glorious sound of the big guns boomed once more. The howitzers pumped their shells, one after the next, into enemy fortifications, splintering trees and pockmarking the hills. Clusters of enemy bodies popped into the air and came down as corpses.

But something had to be done about the mortars raining down on their positions. Dave rallied a handful of Koreans and led them toward a nearby 400-foot ridge. A frontal assault was the only way to take out the enemy. He pressed hard to the out-skirts of the village, but the fifteen KMAGs and the 300 ROKs kept dropping, forcing Dave to retreat and replenish his squad.

After ten desperate attempts to take the ridge, and after los-ing too many soldiers, it became clear that the Chinese circle would not bulge. The refurbished resistance was crumbling. Hundreds were wounded or dead, making it now impossible to defend the perimeter. The enemy mortars continued pummeling their positions, demoralizing the ROKs and the KMAGs. When the heavy artillery ran out of ammunition, what had started as a bad situation suddenly turned critical. The area was lost. Dave scoured the sky, pining for an ammo drop.

Just then, the sound of a familiar engine rose above the roar of battle. It was the Martin B-26 Marauder, a thin-skinned, accident-prone, hard-to-tame medium bomber that had been designed for speed, not low-level strafing. With her short, overloaded wing and powerful twin engines, the B-26 wasn't ideal for loitering. The unlucky gunner, or "flak-bait," manning the .50-caliber machine gun in the Plexiglas nose would be vulnerable at low altitude.

The B-26 performed best up high. In World War II, the bomber had earned an impressive kill streak against Japanese cities. Experienced pilots called her the Invader. But to the un-fortunate novices who crashed her, and many did, she was known as the Widowmaker. Flying dangerously slow through the Korean mountains, the high-speed bomber required a capa-ble docent. But there was no guide in sight.

Dave lunged for the radio. "Can you help us?" he pleaded. The pilot probably could, but not without a Mosquito plane cu-rating the attack. Dave had an idea. *He* would become the Mosquito.

"Fly directly over this village heading north," Dave instructed the pilot. "I'll tell you when to start firing." The pilot consented, leveled off at 100 feet, and dropped his speed to 140 miles per hour, keeping her just above a stall.

Dave saw the Invader growing larger on the horizon as she came up the southern road. The moonlight stenciled a handsome silhouette around its bullet-shaped nose, which had to pass directly over Dave's head before he could give the command. Timing was everything.

Dave kept his finger on the transmitter, waiting for the right moment. A few more seconds lapsed.

"Fire!"

The B-26 opened its mouth and vomited a brilliant stream of .50-caliber slugs at the charging Chinese. A cheer erupted from the ROKs as the front turret blazed a trail of red mist through the enemy.

The celebration, however, was short-lived. In all the commotion, Dave had forgotten one essential thing about the B-26—it had a rear dorsal turret, and it was also firing.

A second line of lightning followed behind the first. The tail gunner, unaware that he was shooting into friendly forces, scattered a constellation of bullets at Dave and his men, who dove for cover in surprise. Nevertheless, the Chinese rifles turned suddenly skyward. The B-26 unfocused the enemy and created enough confusion in the hills to temporarily halt the attack. An ROK soldier clambered up to Dave with encouraging news. A cache of 1,300 artillery rounds had been discovered in a nearby cave.

"Get the guns going!" Dave yelled.

But no one obeyed. Dave repeated his order, which was again met with silence. He started for the cave himself, hoping that if he led the way, a few soldiers would follow. It was no use. The incorrigible ROKs were too shell-shocked to budge.

Frustrated, Dave removed his .45 from its leather holster and smashed two Koreans across their heads. It did the trick. The men spurred into action. The squad retrieved the ammo, loaded the big guns, and launched a salvo of fire that lasted the rest of the night.

Over Korea, the Land of the Morning Calm, dawn came and with it came the sound of a Mosquito buzzing above the resistance. Dave scrambled to the radio. For the next three hours, he worked in tandem with the pilot to coordinate a myriad of air strikes against the enemy troops.

The incoming rifle fire intensified, and to make matters worse, Chinese snipers were picking off Dave's gunners. He passed the microphone to his operations assistant, grabbed two grenades, and crawled away from the jeep, edging along walls of nearby buildings until one of the snipers came into view. Dave pulled the pin and threw it at the sniper. The frag exploded. He chucked another one for good measure and crept back to the jeep.

This time he grabbed as many grenades as he could carry. He moved stealthily between buildings, crawling at times on his belly. He pitched a grenade, then another. When he had taken out a half-dozen snipers, Dave returned to the jeep, depleted.

His operations assistant was sitting motionless by the vehicle. Blood was spurting from the bullet hole in his forehead. Dave took the radio transmitter from his hand and continued calling out targets for the Mosquitos flying above, crop-dusting the ravine with bombs.

But the Chinese kept coming. With no regard for their own dead, they climbed over the mutilated bodies in a desperate attempt to secure a stranglehold on the Soyang River valley. They were close enough now for hand-to-hand combat.

Dave lobbed another grenade in their direction and took cover beneath the jeep, his hand in a death grip around the radio transmitter. A sniper tracking his movements teased the jeep with a few staccato shots. Dave tucked his limbs out of sight.

The voice of a Mosquito pilot came in over the airwaves. "A relief column is marching up from the south," the pilot said. "Hold on a little longer!"

Dave was exhausted. He had not eaten or slept since the attack began, and a mixture of blood and sweat pooled in his eyes, clouding his vision but distracting him from the metallic taste in his mouth. Something was wrong with his jaw.

The advancing army swarmed up the road in their olive drab green uniforms, a welcome contrast to the mustard-colored cotton the Chinese were wearing. Dave breathed a sigh of relief until the friendlies began firing in his general direction. The jeep pinged and a few of the ROKs collapsed. The allies seemed to be aiming at them.

Dave wiped his eyes and glanced again. It was a trick. A battalion of Chinese soldiers had disguised themselves in Korean garb to infiltrate from the south. Dave's division was now surrounded on all sides, and there was only one thing left to do.

Move!

Dave rolled from beneath the jeep and, with a thermite grenade, set fire to the heavy radio boxes in the trailer. The cumbersome equipment would be useless in a high-speed escape. He grabbed an armful of weapons and stashed them in another vehicle: a cluster of grenades and an M-1 rifle and a .30-caliber, semiautomatic carbine. He also chucked two portable AN/ARC-7 radios, knowing that air support would be essential for the bugout. Dave strapped his camera around his neck and jumped into the jeep, along with his Korean interpreter, who was wide-eyed with fear.

The dwindling KMAGs and ROKs followed Dave's lead. A mortar landed on one of the vehicles, bursting into a balloon of glass and shrapnel.

Dave maneuvered his jeep to the front of the caravan. He wasn't the senior officer, but someone had to spearhead the escape. Just where in the hell would he lead them? It didn't really matter. If they could get south, through the Chinese, they'd be safe.

The city of Inje was a few miles away, but rushing up Route 24 was a frenzied mob of enemy soldiers—hundreds of Chinese disguised in Korean uniforms. They were charging in what the Chinese called *renhai zhanshi*, or "human waves." It functioned like a human conveyor belt, sending waves of soldiers, one after the next.

With a reckless abandon known only to those about to die, the newfangled leader barked out a few orders and rested his boot—still tightly laced—on the accelerator. Nineteen jeeps revved behind him as Dave white-knuckled the steering wheel and braced for the bloodbath.

There would be no turning back.

The moments that followed would later be lionized as "Dave MacArthur's one-man war." But as his vehicle forged ahead, plowing through the Chinese bodies, Dave knew only one thing for sure: he had been captured once before and, by God, he wasn't going to be captured again.

CHAPTER 2
KEEP 'EM FLYING!

Davey Warren MacArthur was born into a world between
wars.

The MacArthurs welcomed their first of three sons into the
suburban community of Brighton, Massachusetts, on August
31, 1924. On that very day, Adolf Hitler was busy dictating his
book *Mein Kampf* while serving a prison sentence in Munich,
Germany, for attempting to overthrow the Weimar Republic.
His Beer Hall Putsch coup had failed, but the thirty-five-year-
old had loftier goals in the works. Upon his release from
prison, the future Führer planned to marshal Nazi support, conquer Europe, and restore to Germany the honor she had lost
five years earlier, when the Treaty of Versailles had ended the
"war to end all wars."

Hitler's autobiographical book soon became a bestseller. He
was appointed chancellor of Germany in 1933 and assembled
his own private militia. By 1938, Hitler's face appeared on the
cover as *Time* magazine's "Man of the Year." He had become,

as the article touted, the "greatest threatening force that the de-mocratic, freedom-loving world faces today."

When Dave was five years old, the stock market crashed, ushering in a decade of poverty. Like many American families, the MacArthurs were forced to tighten their belts. In 1930, one year into the Great Depression, Dave's father, a Methodist minister named Vaughn H. MacArthur, reconnected with the Army as a Reserve Chaplain.

After Vaughn was called into active duty in 1939, he and his wife, Dorothy, transplanted the New England family often. Dave never remained in any one place for more than about three months. He finished elementary education in Ohio and gradu-ated high school some two thousand miles away in Lompoc, California—the ninth high school he had attended.

The Second World War had been raging for three years when Dave exchanged the West Coast for the East Coast and enrolled at South Carolina's Clemson A&M College, a mili-tary institute and proud home of the Tigers.

Athleticism ran in the MacArthur family. Like his brother Gene, who would begin his college football career at Vander-bilt as the third-string freshman center two years later, Dave had a lean physique and legs as lively as his imagination. He excelled equally at football, basketball, and baseball. His toothy smile was broad, as were his shoulders. And in his eyes, which were as piercing and playful as the Pacific Ocean, one could see a good-natured boy becoming a man. Years later, long after Dave's adolescent jaw had squared beneath a hand-some wave of thick red hair, people would say he favored Ted Kennedy. The comparison was apt, reminding Dave of a chance meeting in a Boston bar shortly after the war among Dave, his buddy, and Ted's two older brothers. What Dave lacked in temper, his mouth made up for in spades. When hot-headed Bobby brushed up against cool-natured Dave, sparks ignited into something of a brawl before the young men were thrown out of the bar and the fight fizzled.

Clemson took its ROTC program as seriously as it took its football, producing a covey of officers for the U.S. Army. Graduates from the Reserve Officers' Training Corps were promised bright careers.

"Repeat after me," Dave was instructed at his freshman orientation, his hand dutifully raised as the class took the oath. "You're now privates in the United States Army, and when you graduate, you'll become second lieutenants." Before he knew it, Dave had sworn his allegiance to the Infantry. He would receive a monthly stipend of twenty-one dollars while in school before being called into active duty.

But the eighteen-year-old had no interest in joining the Infantry. He wanted to be an aviator, entering the Second World War from above, not below. He imagined his wings, outfitted with weapons, swooping down upon the Nazis, depositing an arsenal of explosions, then disappearing in a wake of coiling smoke. That was the dream.

Dave wouldn't stay at Clemson long enough to become a Tiger. He told the ROTC that he wanted to be a pilot, but as a college freshman, he didn't have much of a bargaining position.

In fall 1942, only a few weeks into Dave's studies, Chaplain Vaughn MacArthur gave his son a heads-up that an Aviation Cadet Selection Board would be near Clemson. Dave and twenty-five of his buddies went down to interview, passed their mental and physical examinations, and within a week were admitted into the Army Air Forces' Air Corps.

The enlistment brochure, designed to recruit promising aviators, fueled Dave's determination to become a pilot.

*We are in the midst of the most momentous war of
modern times. A coalition of powerful and ruthless enemies seeks not only to overwhelm us but to annihilate
our institutions and our civilization. They have struck
with suddenness and with all the force at their command,
and have shown that it is their aim to conquer swiftly
and completely.*

*Therefore, we have no time to lose. We must surpass
them in both strength and speed of attack. We must press
them back behind their own borders and there defeat
them so decisively that they can never again attempt to
impose their wills and their ways of life on a people who
cherish liberty above all things; a people always willing
to lay down their lives to preserve their freedom.*

*The United States is now engaged in the greatest
aircraft production program ever undertaken by any
country. That program, however, can be translated into
air supremacy only if we can muster the qualified man
power to keep our planes flying. And the source of this
man power lies in the youth of the land—they are the
men who will 'Keep 'em Flying!'*

*Youth alone has the physical fitness, the mental alert-
ness, the personal daring to meet the acid test for air
crews of high-powered military aircraft.*

*Our Nation's future depends upon command of the air.
The future of freedom and liberty everywhere is in the
hands of our youth.*

Clemson didn't take kindly to the news. The would-be pilots
were out of the Air Corps. "You're in the Infantry and you're
going to stay in the Infantry," Clemson said. The college
kicked Dave out of class for violating his ROTC contract and
sent him to Fort Benning, Georgia, the "Home of the Infantry,"

to continue his training as a private in the U.S. Army. The crest-fallen freshman had run out of options, and he turned once again to the only man he trusted to come to his rescue.

When Vaughn MacArthur hung up the phone, Clemson had a sudden change of heart. The college granted Dave permission to disenroll, and within days the fledgling flyboy had his orders from the Air Corps. By November, Dave was in Miami Beach, Florida, to begin his initial aviation cadet training.

He knew it would be grueling. The command of an aircraft was Dave's sole responsibility. He would have to train longer than the navigators and other members of the air crew. During the thirty-six weeks of instruction, he had to demonstrate quick reflexes, perfect physical coordination, an aptitude for me-chanics, and the ability to make rapid decisions under duress. Beyond those demands, he had to master the fundamentals of general military training before he could progress to primary, basic, and advanced flying.

It was a pretty good gig. Dave would receive fifty dollars per month during his preflight training, then seventy-five after his appointment as an Aviation Cadet. He would be furnished with living quarters, medical care, uniforms, and a $10,000 life insurance policy courtesy of the U.S. government. After he completed his courses, Dave would be commissioned as a sec-ond lieutenant and assigned to active duty with the Army Air Forces.

Eventually, he would have to select a particular field of spe-cialization. Which plane would he fly? Twin-engine or single? What kind of mission did he desire? To bomb or pursue? These were daunting decisions for the teenaged cadets. They not only steered the trajectories of their careers, but they also deter-mined who would live and who would die.

A world of difference distinguished the fighter pilot from

the bomber pilot. The fighter pilot was naturally competitive, athletic, often impulsive, and never content to fly straight and level. At full throttle, he must be comfortable aiming his engine straight down at the ground to pursue a target. Adrenaline, fear, excitement, anxiety—these all had to be channeled. Split-second decisions had to be made at terrific speeds. Life depended on it.

Flight instructors could spot a fighter pilot from across the room. He was usually the leader of the pack, the captain of the team who rallied his mates to victory. The worse the odds, the better he performed. Being a fighter was a high-risk, high-demand job, and not every cadet was cut out for it. Robin Olds, the legendary P-38 Lightning pilot, summarized it well: "Bombers drop bombs. Fighter pilots fight. . . . The bomber guys have a committee to tell them what, when, and where. . . . The fighter pilot is driver, navigator, gunner, bombardier, and flight engineer wrapped into one tense, high-strung package."

Dave MacArthur was born to be a fighter.

After completing six weeks of basic training in Miami Beach, Dave traveled to Furman University in Greenville, South Carolina. The college detachment school was only twenty-five miles from Clemson, where his journey had started. The destination was worth the detour.

Dave's first aircraft was the Piper J-3 Cub, a top-winged hand-me-down donated by civilian flight instructors. By the look of her wheels, the trainer had seen more than a few rough landings. Her threadbare seats and faded instrument panel suggested years of abuse. Ten hours of flight time were required in the Cub if Dave wanted to live up to the motto of the U.S. aviation cadet training program: *Ut Viri Volent*, "That Men May Fly." He had to tame the old bird, but success wasn't guaranteed. Furman was the first phase in which the Army Air Corps began eliminating people. Three guys in his cadre had already failed their flight examinations and washed out of the program.

Aviation was still an industry in its infancy. Only forty-four years earlier, Wilbur and Orville Wright, bicycle mechanics from Ohio, had cast their daring vision. Wilbur immortalized it in a letter written to the Smithsonian Institute in Washington, DC: "I believe that simple flight at last is possible to man."

He was right. Four years later, their dream became a reality. On December 17, 1903, along the coastal sands of Kitty Hawk—only five hundred miles to Dave's immediate east—the first heavier-than-air machine distanced itself from the North Carolina dunes. The Wright brothers sustained a remarkable flight that day spanning 120 feet, less than half the distance of a football field.

Five years later, the Wright brothers invented a biplane with double elevators and a rudder in the front. It weighed four hundred pounds and could produce speeds up to forty miles per hour after being launched from its monorail.

A newly created aeronautics division of the U.S. Army took an interest in the aircraft. In 1909, the Office of the Chief Signal Officer approved the biplane for military use, and two years later, Congress allocated $125,000 for the advancement of aviation. The first Acro Squadron came together in 1913. Its fleet began operations in Mexico with eight planes, sixteen officers, and seventy-seven enlisted men. When the United States declared war on Germany for the first time, the Army Aviation fleet grew to fifty-five planes and thirty-five pilots.

But everything changed in 1935 with the creation of the GHQ Air Force. For the first time, pilots were trained with standardized instruction across the country. Even more significantly, the U.S. Army Air Corps came under a single command. By 1941, the Army Air Forces was established, and the rules of modern warfare would never be the same again.

*   *   *

Dave climbed into the front seat of the Piper J-3 Cub and settled into the controls. The stick was smooth as stone, polished
by the hands of a hundred pilots before him. The instructor
cranked the propeller, took his seat, and when the oil pressure
was up and the altimeter set, he taxied the plane to the line.
Like every first-time pilot, Dave was about to experience the
exhilaration of pulling away from the ground. He would soon
be looking down at the clouds instead of up at them. The future
was waiting at the end of the runway, and Dave raised his gaze
to greet it.

If only his father could see him now.

CHAPTER 3
HATBOX FIELD

Summer 1943

If hell had a temperature, Dave discovered it in Texas.

After his debut flight in the J-3 Cub, he traveled to the San Antonio Aviation Cadet Center. The foothills of South Carolina's Blue Ridge Mountains were behind, and his fair skin now soared through the sticky air above the fourth hottest city in the United States. San Antonio was nestled along the southern end of the Texas Triangle and borrowed its name from thirteenth-century Franciscan missionary Anthony of Padua. He was commonly known as the saint of lost things.

Dave lost sight of the ground as his instructor pitched his plane to thirty-eight thousand feet. The summer-scorched city vanished below as the sky opened into a hazy yellow horizon. Everything was jaundiced by the Texan sun. Dave had to pass his pressure chamber examination. He had to demonstrate his ability to function at high altitude. But the oxygen mask carved deep lines into his sunburned cheeks, distracting him.

The San Antonio flight training included endless hours of

coursework, seven days a week. "Boy, they do not give you time to think," Dave wrote his parents.

> My first class was Naval Identification. We have it
> two hours a day. It is an 8 hour course in which we learn
> to identify two hundred ships. It is really something. I
> guess it is going to be really hard. Next we went to math.
> We have 20 hours of that, and that is also hard. Enough
> said about that. The last thing we had this morning was
> code. It is going to be really tough. We have to learn
> code, plus the phonetic alphabet. It is really something.
> He fed us seven letters the first 45 minute period. We got
> them fast, too.

Physics, and maps and charts rounded out Dave's coursework. They were essential. No pilot could afford to get lost when flying over enemy territory.

When Dave wasn't studying, he marched. When he wasn't marching, he drilled. When he wasn't drilling, he pushed through hours of physical training in the sweltering heat. "When the sun is out," he wrote his dad in June, "it really is scorching. I am now missing part of my face where the sun has blistered it, and the skin has peeled."

He stole a few precious seconds during class to scribble a few more words to his folks.

> This writing right now is hard. My muscles are
> fatigued from Physical Training. I have another burn
> coming on. Here is an idea of how hot it is. We started
> picking up the rocks on the field. The rocks were so hot
> that they burned a fellow's hand. Then we started the ex-
> ercises. Nine tenths of them are on the ground. It is really
> hot there. The temperature is over 100 right now. My

skin must be tough, for I spent over an hour in the sun
with no shirt on.

The Army fed him well, but Dave was burning too many
calories. His ravenous appetite was impossible to appease,
gnawing through the nights.

For the first time in his life, Dave finally had some money,
but because of the fast-paced curriculum, the men had little
personal time to spend it. "I got your letter this morning," he
wrote his dad.

I would have gotten it sooner, but we had open post. I
went into San Antonio to get some stuff. I looked around
for something to get for you, but I guess I am not so hot at
that. I couldn't find a thing. I feel a lot better now that I
have some money. It is not much, but it is enough for me.

In San Antonio, Dave also got his first taste of the Army's
gig-slip demerit system. "They are getting pretty strict around
here," he wrote. "Today, they pulled a kangaroo inspection.
That is when you are gigged even if you have nothing wrong.
It was my first and last gig in this place (I hope), so I am not
worried." One hundred of his fellow trainees had already
washed out. Some 250 men had yet to be classified. But Dave
was still holding on. Soon, he'd have good news to share with
his dad.

"I'm finally getting the thing I have wanted and sweated
four months for," he wrote. "I am getting my cadet issue."

July took its time, and August seemed everlasting. But when
summer finally yielded to fall, Dave boarded a train for single-
engine flight school in Muskogee, Oklahoma. He had sweated

enough for one lifetime, maybe even two, and was ready for the cooler climate promised by the Midwest autumn.

"Dear Folks," he wrote on September 2. "I have at last hit the half-way mark on my way towards my ultimate goal. It is really a good deal in many ways. I sure hope I like it."

Two days earlier, at 8:00 A.M. on August 31, Dave and the other cadets had left their pre-flight training and were loaded into "the oldest day coaches" Dave had ever seen. "They were terrible," he said, hardly a fitting way to spend his nineteenth birthday.

To top that off, we had a lieutenant who made us treat them as though they were made of gold. It was a lucky thing the engine was an air burner. There was no ventilation, so we had to keep the windows open. A coal burner would have made that almost impossible. As for the route we followed, there's only one explanation for that. By the way we traveled, it was assumed that the engineer was dodging torpedoes.

By the time they arrived in Muskogee, the exhausted cadets had traveled nearly the full length of Texas and most of the Sooner State. None of them got any sleep, huddled together three to a seat in the crowded chair car. Roll was called, and Dave then boarded a bus.

It was 7:00 A.M. on September 1. For the first time in sixty days, Dave was greeted with rain. His new home was nestled in the Arkansas River Valley, and to the road-weary men it might as well have been a suburb of Paradise. They rode over a winding road and along the remains of a golf course, relishing the drive to the school, where an imposing gate greeted them. It was the first thing Dave saw. The elaborate gate of the Muskogee Spartan School of Aeronautics looked like it belonged to a fancy country club.

Before the war, the school was the largest civilian command field in existence for training U.S. pilots. In the year of Dave's birth, its famous airstrip, Hatbox Field, had attained some historic notoriety when the legendary Douglas aircraft graced its runway at the tail end of the 1924 Around the World Flight— the first aerial circumnavigation of the world. Now, Hatbox Field trained mostly newbies. "Men paid good money to come here," Dave wrote about the civilian outfit. "Things are fixed up pretty swell. Our barracks are big and cool. They all are done in a fine mahogany finish and are really swell. There is plenty of lights, too. Compared to pre-flight, this place is heaven. It really is nice compared to the rest of the places I have been. As it is now, every month we get 127 dollars from the government, and then we pay the school. This training we get here is worth 2,100 dollars. I guess we are quite an investment now." Being assigned to a civilian flight school was a choice assignment, and Dave knew it.

The food was anything but spartan. Quality training required quality nourishment, and Dave's appetite finally met its equal. He couldn't say enough about the chow. "We walk into the mess hall, what is nothing more than a high-class restaurant, and grab anything you want." Compared to the GI Army food he was accustomed to, it was "really something."

After a brief Army physical, Dave collected his helmet and goggles. He was plenty messy and used every one of the fifteen minutes allocated for a shave and shower. A few students nodded off during the commanding officer's lecture, but the CO couldn't blame them. The pilots would need all the sleep they could snatch before their training began.

Within two weeks, Dave discovered a way to mix business with some pleasure in Oklahoma. He attended a local Methodist church, met an attractive young lady from the choir, and took her on a date.

His new aircraft was equally easy on the eyes. When Dave saw the Fairchild PT-19, with her bright blue fuselage, canary-colored wings, and American flag–striped rudder, it was love at first sight. The two-seater, low-wing monoplane bore the number *99* on either side. When seen from below, she spelled "U.S. Army" in large black letters that were painted between red, white, and blue circled stars. The ship carried herself, and her pilot, patriotically.

The Fairchild cruised at a confident 120 miles per hour. When asked, she could punch a faster hole through the clouds and better resemble the combat birds every cadet dreamed of flying. At Hatbox Field, Dave would get sixty-five hours in the "Cradle of Heroes," as she was dotingly described. The upgrade, however, came with a catch—the PT-19 had an open cockpit.

For the cadets in the barracks below, the weather was tolerable and offset by heaters. For the pilots soaring above, however, the frigid winds were inhospitable. The chill always found a way to penetrate the fleece lining of Dave's flight suit. It spilled out around his neck and over his shoulders, but not even the six inches of insulation could keep him warm. Dave had traded one extreme for the other, and he shivered in the cockpit, his cheeks once again blistered, this time by frostbite. He could use some of that Texas heat he once so eagerly escaped.

Flying solo required smooth landings, but these were hard to come by, especially with numb fingers. Dave's feet trembled against his pedals, which yawed the ship side to side like the needle of a compass trying to find its north. Some of his classmates scored an impressive fifteen to twenty touchdowns in a single day. Dave had totaled only seven.

After a bouncy attempt to stick his ninth landing, Dave's instructor ran out of patience. "Get me out of this damned thing before you kill me!"

Dave landed the plane and his instructor jumped out, grateful to be on the ground. The chance to solo suddenly presented itself, and Dave took full advantage, turning back to the runway, calming himself, and then gunning the plane.

Without a second body in the cockpit, the PT-19 lifted quickly. "It was one of the bumpiest rides I have had," he wrote his parents. "The wind was really bad." Dave shot one landing then rose, banked, and lined up for another.

The ship, like her pilot, had performed well enough, but it wasn't over yet. Before Dave's wheels kissed the runway, something unexpected came careening across the field.

A bus.

The vehicle swerved onto the tarmac, unaware that a plane was trying to land. Dave panicked, put full power on the plane, and pulled up. The engine screamed for airspeed. Math began to matter more than ever. Time seemed to stop as the scene unfolded in Dave's mind. First, there would be the collision. Then the explosion. His plywood wings would snap, and the fuselage would crumble. A few minutes later, an ambulance would race down the runway, following the trail of fire and smoke to Dave's smoldering corpse.

CHAPTER 4
THE NORTH POLE

Dave jerked himself back to reality as his wings found enough lift to rise above the bus. He had avoided the collision, and out of his mouth came the sound of something between victory and disbelief. This pilot's time had not yet come.

He banked around for a second landing, double-checked the runway to make sure it was clear, then touched down. Dave's first solo flight was under his belt, and he basked in the thought that he was one step closer to combat.

The rest of the time in Muskogee Dave spent performing maneuvers. He threw his plane into lazy eights, pylon eights, wingovers, spins, and chandelles. "I did just about everything in the books," he wrote his father. "You cannot imagine how much fun it is to get up there alone on a solo flight at about four or five thousand feet and just throw the ship all over the sky."

The rush of looping, the suspense of stalling, the miraculous joy of sweet recovery—each of these rewarded the nineteen-year-old with a new sensation. Even gravity, that unbreakable law of old, couldn't hold Dave down. When leveling out of a sharp vertical rise, he was suddenly weightless, as though his body could float into space if not firmly tethered to his seat.

At times, a pilot's experience could edge on the spiritual. The runway became a nave; the clouds, a cathedral. When rising into an evening sunset, the line between earth and heaven vanished. Rainbows, when seen from the ground, are always cut in half. But in the cockpit, their spectrums appear fully orbed, an effervescent color wheel painted for the few lucky pilots who see them. In the sky, Dave became fully alive. But the stress of training, combined with his near brush with disaster, soon took its toll. The kid uncoiled his emotions in a letter addressed to an Army chaplain who was familiar with both sides of death's door. "Dear Dad," he wrote,

As far as these letters go, telling you of my feelings, they are more of a morale builder for me. It is a lot easier to take if you can get it off your chest. You are the nearest chaplain I can go to. I guess that is one advantage of your being a chaplain. I can go to my family as well as the chaplain all at the same time.

November 1943, Coffeyville, Kansas

Dave was soon bound for the barren lands of Kansas. After seventy hours of flight time in Oklahoma, he was ready for the Stearman PT-13, a docile dream to fly. With its open cockpit, the pilot didn't even need to check his airspeed; the sound of the wind whistling through the cables said it all.

Other than its yellow-tipped black propeller, the biplane had the same paint scheme as his previous plane. He couldn't wait to get his new ship sailing. Dave completed the cockpit checklist, idled his engine, and gave a cheerful thumbs-up to the crew chief, who removed the wheel chocks. The Stearman's nose was angled high. Even with the seat raised, it was difficult to see over. Like a rhinoceros, cursed with eyes on the sides of her head, the plane had to zigzag left and right to see straight on when taxiing. This sailboat motion, or S-turn, was made possible by the rotation of the rear wheel, which Dave reminded himself to lock before takeoff.

He reached the runway and pressed the balls of his feet against the rudder pedals, activating the brakes. He lowered his seat to be eye level with the windscreen, slid his heels to the deck, and gave the rudder one final wiggle. He was cleared for takeoff.

He smoothly applied the throttle, and the biplane was in the air. It behaved brilliantly. There was more wing to this plane, more control surface. At 160 miles per hour, it responded almost instinctively to the pilot's demands. When asked to dive, the bird swiftly bowed its beak. When commanded to bank, only the slightest flick of the stick raised her wing in sudden salute. Flying in tight formation was trickier, a surefire lesson in stress management. Only thirty-six inches separated Dave's wings from his buddies'. One absent-minded mistake, one mistimed maneuver, and in a split second he could take out his squad, make a quick end of himself, and cost the government thousands of dollars in flight equipment.

In December, Dave wrote a sobering letter for his father's eyes only.

I don't suppose that mother will ever see this letter, so I think I had better tell you one or two things. My bunkmate is in the horizontal from one wreck, and the fellow in our flight in the next barracks was shipped home, or what was left of him, while the instructor I was assigned to temporarily was also shipped home. Our squadron has had a lot of tough luck, and it makes a fellow begin to realize what this is all about. I am never going to mention this to mother, but I thought that I had better say something to you. I do not know how much info they let us get out so I keep my mouth shut. We are flying in some pretty cold weather, though, and things are bound to happen, so I want to keep on the safe side. They really make something out of you around here. Flying as we are, in all sorts of conditions, it either makes you a good pilot or a dead one.

December

The bitter weather was chasing at Dave's tail.

"When I first came here," he wrote his mother at the end of the month, "I thought that Kansas was a state. Now I know they made a mistake, for this is the North Pole. Boy, this place is cold. You go up 5,000 feet in a plane and it is 30 or 40 below. I have been wearing every piece of clothing I can get plus a flying suit and I still freeze myself solid."

His next aircraft, the Vultee BT-13, had a closed canopy held together by thin metal strips that curved around the pilot's head in the shape of the letter *A*. It was warmly welcomed. The students called her the "Vibrator," and for good reason. With 450 horses inside her Pratt & Whitney engine, the two-position propeller produced a high-pitch tremor that violently shook the canopy. When stalling out, the plane more or less had a seizure.

Compared to the lightweight biplane, the Vibrator wallowed through her aerobatics class "like an old lady," as Dave described it to his brother Charlie. The plane was also less responsive, noisier, and even incorrigible when her crank-and-cable flaps refused to cooperate. She could be greedy, too, demanding a generous landing speed of eighty miles per hour.

"Still and all," Dave wrote, "they are a hotter ship than the PT-19." The plane was a rite of passage. Almost every U.S. pilot in World War II trained in the BT-13. Many pilots sent their first air-to-ground radio transmission from her cockpit. Her bigger engine was worth the diminished maneuverability. She could catapult two tons of metal at a swift 182 miles per hour. When flying at low altitude, the landscape became a blur, ruining pilots for slower wings. Dave reasoned its cavernous, 120-gallon fuel tank could take him all the way to Nashville or Louisville.

He'd soon have a chance to test his theory during his first cross-country, long-distance flight. "Well," Dave wrote his parents, "your little boy is beginning to travel a little. Our x-c flights will have from 200 to 400 miles in length. It takes about two hours at the most to cover over 200 miles. I can remember when 200 miles meant half a day on the road."

For three hundred years, the world had been shrinking. In the eighteenth century, New Englanders traveled primarily by foot, horse, stage wagon, and boat. Fur trappers could push into the continent's interior, but they were forced to navigate uncharted rivers and Native American trails.

In the nineteenth century, as forts became towns and "gold fever" mobilized the population westward, the blazing of roads cut travel time significantly. So did the newly laid railway tracks that promised to sinew the country together. In 1869, the East Coast touched the West Coast when Leland Stanford hammered the last nail, the "golden spike," into the first transcontinental railroad. The width of America, once a half-year journey, could then be traversed in only seven days. Seventy-five years later, in 1943, Dave could cover a trip from Kansas to Boston in three hours.

"It is hard to realize," he wrote, "that that much distance can be covered in such a short time. It is a thrill that only 450 horses in front of you can give."

The thrill of flying had a flip side.

"You sweat blood in this business," Dave wrote his dad. "These fellows, I'm included, wash up at night in all sorts of situations. I have heard fellows curse instructors in their sleep. I have cracked up and made every single mistake there is in the book, in my sleep."

Soon, Dave would have less sleep to worry about as he took up night flying, an exhilarating but disorienting experience in the dark. "If there has been anything difficult in flying up to now," he admitted, "it was just kids' stuff."

In the absence of visibility, the pilots couldn't judge distance. Dave had to blindly trust his airspeed indicator, compass, altimeter, bank and turn indicators, and tachometer. These instruments were like a sixth sense that had to be developed in the dark.

"When you are in the plane," Dave described to his brother Gene, "there is a panel full of flowing instruments. You cannot see much, and you are looking for other planes all the time. With the canopy closed, you pass over a lighted city and see it reflected above you in the canopy. You get the impression you are upside down."

When passing through a cloud, Dave's canopy suddenly turned a crimson color, the reflection of the plane's navigation lights. It was even more surreal with the lights off. Everything went blue because of the engine exhaust, which produced a long streak of burning gasoline all the way to the tail.

"With all of that," Dave explained, "you're pretty much confused. You really have to keep your head."

By the time pilots returned to their sack shacks around

3:00 A.M., they were usually dead-tired and sore all over. "Boy, I am not complaining," Dave wrote, "but they are shoving this stuff at us too fast. It is tough to take it all in such a big dose."

Frostbite became an ever-present enemy, one that grounded nearly half a dozen pilots at any given time. At night, the temperature in the cockpit dropped to twenty degrees below zero, and the parachutes beneath the pilot became rock hard.

In the event of an emergency, it was up to the cadet to decide whether to stick or bail—a tough decision to make on only four hours of sleep. "When you figure how close you can come and still make it," Dave wrote his dad, "you begin to wonder."

One experience gave Dave legitimate cause for wonder. He had been flying for two hours in the dark and had shot eight landings. To rush things along, his instructor gave him credit for seventeen landings and called him in.

As Dave made his final approach, the runway, once lined with lights, went totally dark. He was easing back on the throttle but couldn't see a thing. No horizon, no moon or stars, just jet-black midnight above, below, and beyond. It was as if he were landing with his eyes closed. Dave didn't know where his wheels would touch, but he descended anyway, trusting that his wings were somewhat parallel to the runway.

"I brought it in," Dave wrote, and stuck a fine landing, too. But "it is a funny feeling to land on something you can't see."

The Link flight simulator was much safer. The Army Air Forces Pilot School required cadets to spend eight hours training in its mock cockpit. First developed in the 1930s, the simulators rose to popularity among World War II aviation schools like the one at Coffeyville. Dave took the simulator as he took to the skies—with equal seriousness. The aviator needed all the training he could get.

In the air, pilots had to navigate a series of turns and maneuvers without varying more than fifty feet in altitude. Each variance cost the cadets twenty-five cents, the equivalent of about four dollars today. Dave was determined to avoid the debt, and he also had to keep his mouth in check, because the price for cursing on cross-country buddy rides was steep.

"Boy, it really gets rich up there at times," he explained in a letter to his dad. "It is a $10,000 fine and 20 years in jail for cussing over the radio. When these buddy rides start every day, some of them forget to switch from radio to interphone. You can hear the poor guy in the tower plead with the fellows to switch to interphone."

One December day, Dave's mouth got him into a different kind of trouble. He was training in the simulator, about to complete his quota, when his sergeant approached and attempted to speed the process along. But Dave would have none of it. He held up his hand in defiance and continued the simulation.

"That's the wrong attitude, cadet," the sergeant warned.

Dave could not be dissuaded. "I don't give a hoot about my attitude," he quipped. "I want to get Links done with, and I'm not going to be screwed out of it just because I lack twenty-five minutes more."

The sergeant gave in and let him finish. Within the year, it would become obvious that Dave's stubborn optimism proved to be the right attitude after all. He had inherited it from his father, and without it, the cadet could not have survived the perils ahead.

Dave didn't make it home for Christmas that year, so his mother drove the five hundred miles from Camp Polk in Leesville, Louisiana, to pay her firstborn a visit.

Dorothy MacArthur arrived on New Year's Eve, 1943. She was a woman of smallish stature but scrappy and full of fortitude. Her rounded spectacles were not much bigger than her eyes, and in the absence of a smile they produced a bookish look to her face. In reality, though, Dorothy was quite the extrovert, a trait Dave inherited, along with her sense of adventure.

It had been nearly a year since he had taken a full day off. That was a problem Dorothy was determined to remedy. When Dave got off at 4:00 P.M. that afternoon, the motels of Coffeyville were already full. Mother and son set out for Tulsa, some seventy miles away, and booked a cabin on the outskirts of town. The weekend rate was four dollars a night, a small price to pay for some much-needed R&R. The exhausted teenager had trained for weeks on no more than a few hours of sleep each night, constantly torn between the exhilaration of flight and the underlying fear of inadequacy. That Saturday morning, however, sheltered within the womb of the woods, Dave slept until 11:00.

The whole next day was spent in restaurants and the movie theater. "I had a swell time," he wrote his father. "We ate, saw a show, ate and saw a show, and ate again. I made up for the food they give us here!"

Dorothy's visit had refueled her pilot, but Dave sure missed his dad, especially after seeing a visiting colonel spending time with his son. "It kind of made me wish you were here now," Dave wrote, wistfully.

In early January 1944, he hoped to visit his dad at Camp Polk, but Dave's plan was blown to bits when a high-pressure front concocted a blizzard. A cruel layer of snow accumulated on the runway. Across the airfield, Dave could see the wings of a heavy bomber, the B-24 Liberator. Its four engines had been silent for nearly a week, grounded by the inclement weather. "It is really a big hunk of plane," he wrote his dad. "It has enough armament on it to sink a battleship. It is, in fact, a flying battleship."

Some two years prior, in November 1942, Louis Zamperini, an Olympic runner turned pilot, had crashed into the Pacific Ocean aboard one of these heavy bombers. Dave eyed the giant every day, watching it sleep beneath its blanket of white. "Boy, it was really a neat plane," he admitted, but "that stuff is not for me."

Dave had smaller wings in mind, preferably the sleeker ones attached to the North American AT-6 Texan. When one of them landed on the runway in Coffeyville, Dave felt an instant connection. "This one carries all of the guns," he wrote to his dad. "It is quite a job. I sure want to fly one if that is at all possible."

Outfitted with four .30-caliber guns and a healthy rack of bombs, the AT-6 was every bit the flying arsenal Dave desired. To him, she was "the sweetest flying ship in the world." In the Korean War, he would affectionately call her the Mosquito, but at the moment, the trainer's best years were still ahead.

The time had come for Dave to decide between twin- or single-engine flight school. He was on the fence and needed his dad's advice. "I think I know," he wrote, "but I am not enough of a flyer yet to really know. Maybe I will flip a coin. Personally, I will fly anything with wings and an engine. But the hotter the better."

The P-38 Lightning was as hot as they came, "but only the hot ones get that," Dave wrote despairingly. "The rest are dumped in the flying-box-cars."

A war room stationed at Coffeyville kept pilots informed of Nazi activity, and there was never a shortage of topics to discuss. The world had been at war for four years.

Hitler's plans had finally come to fruition. On September 3, 1939, Great Britain and France declared war on Germany when the dictator violated the Munich Pact and invaded Poland. At the time, the Polish army consisted of five hundred thousand soldiers, no match for the Führer's one and a half million men.

The Polish skies swarmed with German aircraft. Among the most notorious in the Luftwaffe was the Junkers Ju 52. The rectangular plane was instantly recognizable and nicknamed the "Flying Toolshed" because of her distinctive, corrugated skin that resembled a barn. She had been designed originally as a seaplane, but over the 1930s she evolved to dry land, sprouting legs and landing gear. When two additional lungs were mounted to her wings, "Aunt Ju," as some called her, became a three-motored workhorse—versatile, dependable, and among the Führer's favorites. Hitler named his personal Ju 52 *Immelmann I* after Germany's first ace pilot of World War I, Max Immelmann.

On May 10, 1940, Hitler started his conquest of Western Europe with fighting words.

"The hour has come for the decisive battle for the future of the German nation," he told his troops. "For 300 years the rulers of England and France have made it their aim to prevent any real consolidation of Europe and above all to keep Germany weak and helpless. . . . The fight which begins today will decide the destiny of the German people for 1,000 years."

That day, forty-two Flying Toolsheds dropped their payloads on the heavily fortified garrison of Fort Eben-Emael in Belgium. Some 16,000 troops parachuted into Rotterdam, Leiden, and other cities as 136 German divisions swept through the Low Countries. Within the week, the swastika was raised over Holland. Within the month, 300,000 British and French Allied forces were pushed all the way back to the coast at Dunkirk.

In the United States, Hitler already had a bounty on his head. It came from Samuel Harden Church, a personal friend of Andrew Carnegie, who wrote a curious and widely syndicated piece for the *New York Times* on April 29. Church promised to pay a handsome cash reward of one million dollars "to anyone who will deliver Adolph Hitler, alive, unwounded and unhurt, into the custody of the League of Nations for trial before a high court of justice for his crimes against peace and dignity of the world." The offer expired at the end of May.

What some had mocked as Hitler's Phoney War suddenly turned serious when, on June 21, 1940, France surrendered to Germany at the exact spot and in the exact railroad car where Germany had surrendered by signing the Treaty of Versailles. Great Britain was next in Germany's sights. But the English Channel was a problem for Hitler, one his tanks couldn't solve. It was up to the Luftwaffe to conquer the British Isles.

As London was being bombed, Winston Churchill became Prime Minister of Great Britain. "I felt as if I were walking with Destiny," he later reflected, "and that all my past life had been but a preparation for this hour and for this trial."

The trials were just beginning. By September 1940, the Luftwaffe blitzed London incessantly, dropping bombs almost every day. In eight months, over thirty-nine thousand British lost their lives. Churchill retaliated, relaxing the RAF policy of striking only military targets and instead adopting "area bombing" or "carpet bombing" of civilian population centers in hopes of deflating Nazi morale.

The U.S. Air Force initially opted for strategic bombing made possible by the B-17 Flying Fortress, which resembled a Greyhound bus with wings. When equipped with the much-vaunted Norden bombsight—a telescoped instrument that solved trigonometric problems of when and where to release a payload—the B-17 became the eagle in the skies, flying at high altitude during broad daylight and striking targets with surgical precision. Bombardiers claimed that the new bombsight was so accurate they could drop a bomb into a pickle jar from altitude. But for all its strengths, cloud interference rendered the bombsight useless.

Early on, high-ranking German officers ridiculed the plane,

calling it the "Flying Coffin." And for many pilots, it was just that. A total of 4,752 Flying Fortresses and their crews would be lost. But without the B-17, said Carl A. Spaatz, who commanded the U.S. Army Air Forces in the European Theater, "we might have lost the war."

In 1941, Hitler decided to invade the Soviet Union. He had wildly underestimated Russia's brutal winters and suffered the consequences at Stalingrad and Kursk. There was also trouble for Germany in the south. Prime Minister Benito Mussolini was making plans of his own. He hoped to restore the grandeur of the Roman Empire.

But Mussolini's plan struck a nerve with the Nazis. If Italy fell into the hands of the Allies, and if Western forces established air superiority in the Mediterranean, Germany would be forced to come to the rescue or risk exposing Europe's southern flank. This was precisely what General George C. Marshall, the Army chief of staff, envisioned. The invasion of Italy, he said, "would establish a vacuum in the Mediterranean" that would pull German soldiers away from the English Channel.

Hitler's worries were justified. Mussolini invaded Greece and North Africa, opening a new theater of war. Two years later, the Allies launched a full-scale invasion of Italy, attacking what Churchill called the "soft underbelly of Europe."

In July 1943, Mussolini was deposed and arrested, his corpse hanged upside down in the streets of Milan. By August, the Allies, under the command of the pistol-toting General George S. Patton, seized Sicily.

Germany deployed sixteen divisions to Italy, totaling one and a half million men. Hitler had to relocate some of his troops from the Russian front, which suited Stalin. He also spread his army thinly across northern France, which Britain desperately needed. By September, Germany launched a series of air raids to blockade the Allied progress in Italy. The muddy hills and waterlogged roads seemed to tilt the advantage in favor of the Reich, whose footprint had filled out the boot-shaped peninsula. "I had hoped we were hurling a wildcat onto the shore," said the frustrated Churchill, "but all we got was a stranded whale."

But for all the setbacks, Naples fell into the hands of the Allies by October 1943. The Reich was now fighting a war on three fronts—the British in the north, the Soviets in the east, and the United States and its allies in the south. A perfect storm was gathering, and the weakened German presence near the English Channel meant the Allies were well positioned for the Day of Days, the surprise invasion of Normandy.

But across the Atlantic Ocean, back in the heart of Midwest America, somewhere over the rolling fields of Kansas, a young man was still becoming one with his wings. In time, he hoped to stretch them east, where all the action in Europe was happening. If nothing else became of his life, Dave was determined to show the world, and especially the Germans, the full and furious extent of the American spirit.

CHAPTER 5
ON WINGS OF EAGLES

In June 1941, six months before the Japanese attack on Pearl Harbor, the United States leased two square miles of earth along the Texas-Mexico border. It cost only one dollar. By October, and some eight million dollars later, the Eagle Pass Army Air Field rose up like a phoenix. Three concrete runways, each fifty-five hundred feet in length, stretched through the cactus-covered fields in the shape of a crucifix.

The town of Eagle Pass, originally known as Paso de Aguila, took its name from the hills surrounding the Rio Grande, which resembled an eagle's outstretched wings. For budding pilots like Dave, there was no better place to earn their own wings.

In 1782, when the United States was but a fledgling nation, she claimed the bald eagle as her national symbol. By the twentieth century, the bird had come to symbolize everything the maturing country was becoming—brave, fierce, and fast. With its sharp vision, the eagle can spot a mouse from three hundred feet in the sky, and it is one of the few creatures in existence that can stare directly into the sun.

The Greek philosopher Aristotle was among the first to observe how eagles compel their young "to gaze on the sun before they are feathered." This ability, made possible by translucent eyelids that move horizontally across the eye, weaponizes the eagle with a surefire means of escape. Though rarely chased by other birds of prey, when in trouble the eagle has the ability to pitch into a vertical climb and blind its enemies by flying directly into the sun. Native Americans believed the eagle could do more than even this. To them, the mythological Thunderbird, or Super Eagle, controlled the weather, hurling thunder and lightning from its wings.

The mascot of the Eagle Pass Army Air Field, however, did not exactly live up to Dave's expectations. He was a lethargic longhorn named Tito (*little one* in Spanish). Tito's ancestors, not native to the continent, were imported by Christopher Columbus, who, in 1493, offloaded a small herd of Spanish cattle in Santo Domingo. A few hundred years later, settlers took them from the Greater Antilles to Mexico on the search for gold. In 1836, after the state won its independence, the beasts became known as Texas longhorns.

Tito spent his days on the airfield in bovine bliss, gazing curiously at the Mosquito planes buzzing in the clouds above. As far as steers go, Tito was as gentle as a kitten and occasionally posed for photographs with the cadets. By the time Dave arrived at Eagle Pass for his advanced flight training, the longhorn was a legend.

Sadly, one morning in 1943, Tito was found dead. His flesh was charred to his ribs, his tail stiff as a board. Some tremendous force had reshaped his horns, once as wide and handsome as the gull wing of a Corsair. The culprit came from the clouds. On the previous night, a rare thunderstorm had produced a bolt of lightning that descended upon the Texas airfield. Within seconds, the beast was barbequed.

Dave was among the youngest cadets at Eagle Pass. He was only nineteen years old and lacked a college degree, but none of the pilots were more eager to begin fighter pilot training. However, only one week after his arrival, Dave came to share something in common with the airfield's unfortunate mascot—he, too, was tethered to the ground.

"Dear Folks," he wrote in January 1944. "It looks as if I will never get up in an airplane again. I have been here a week and still have not flown."

On the ground, Dave spent three hours each day in the Link, the training simulator. It had a miniature fuselage that spanned the length of about six feet from tip to tail. When sitting in its tiny cockpit, a full-grown man looked a bit ridiculous; however, the state-of-the-art electronics that generated its motion gave pilots a realistic sensation of flying. "We have a lot of other gadgets to fool with, too," Dave wrote. "If they took them to Coney Island and charged admission, they could make a lot of money in a short time."

One of Dave's first simulators at Eagle Pass went by the name BB Girls. The Link had a toy BB machine gun mounted to its nose and a regular aircraft gunsight through which the pilot aimed. Fifty yards away a paper target, only twenty inches tall and five inches wide, darted from side to side on wheels. Dave shot four hundred rounds of BBs at the rascal, but he scored only thirty hits.

Dave honed his aiming ability during skeet-shooting lessons on Cactus Field. The range instructor told him the secret. "To hit a moving target that is flying through the air," he said, "one must aim at where it is going to be and not at where it is."

The advice would later prove invaluable in the war and also

in the next flight simulator he was tasked to master. It had a full-sized AT-6 cockpit and was designed to simulate aerial combat. The lights went off, a moving screen switched on, and suddenly Dave was in hot pursuit of a bogey. He lined the gunsight ahead of the target and squeezed the trigger. The plane went down.

"It was a lot of fun," Dave admitted. "But what I want is the real thing."

That's what every cadet wanted, the *real thing*. The dedication of the yearbook of Class 44-F reflects this feeling, which was shared by Dave and every pilot who came and went through Eagle Pass:

About fourteen months ago many hundreds of BOYS started out on a long journey. That journey was to take us through the hell of Basic Training where tortured muscles screamed out their protest, through the smart military and academic training of colleges, to Classification, with all its disappointments, and finally through Pre-flight before any actual flying of military type planes.

The eight months of long preparation for the flight training was hard and will long be remembered. We marched and marched until the spirit of doing things in unison was captured. At times we wondered just what the name "Air Corps" meant, for certainly we weren't flying.

But somewhere along the line a metamorphosis took place: we were slowly becoming men. At long last we shipped to Primary, Basic, and then Advanced, all hallmarks in our road to fame.

Yes, the road is long and rough and steep. No one who has gone through would ever recommend it for weaklings. But to you who dare to take up the challenge

we hereby assure you of one fact. There are those who
get through and attain a great privilege, that of being an
Army pilot.

So to those men who will soon follow in our footsteps
we wish the best of luck and hope you will proudly emu-
late our success.

Before long, Dave swelled his chest and said good-bye to
the ground. "Today at last I flew in an AT-6," he wrote his parents. "I went up for the first time and after an hour of dual and
3 landings took it up myself and I even got my wheels situated
in time to put them down again to land. I really shot through
the stages running to solo," he bragged. "I went exceptionally
fast compared to the rest of the crowd."

The AT-6 packed a mighty wallop, even at half-throttle. In
an emergency, Dave could force the throttle past the stop to
reach full power, but it wasn't advisable. The engine, with its
six hundred racing horses, was much too muscular and tended
to rotate the plane instead of the prop. The experienced pilot
knew to counteract the corkscrew effect by adding power and
kicking the rudder to one side, but for inexperienced pilots flying at low altitude, crashing became more likely.

The perils of being a pilot were never far away. "While I
was up there," Dave wrote his dad, "I had a chance to do some
thinking, and I thought about how I was flying about $50,000
worth of airplane, all by myself."

Equally stomach-churning was the drinking water, which
was known to give the cadets the runs. "For the last two days I
have been sicker than a dog," Dave wrote. "It is this water. It is
pure, unadulterated sulfur and a lot of alkali added to make it
more pleasant. I tossed my cookies all day."

But even when Dave was feeling fine, the demanding gunnery training schedule was nauseating. The cadets had to learn

the .30-caliber and the .50-caliber machine guns, and the 20mm cannon, all in six days. There were several hundred parts to memorize and never enough time to study. For sixteen to eighteen hours a day, Dave agonized through the coursework, soaking the sponge of his mind until it was fully saturated. On Valentine's Day, a break finally came, and Dave got a few hours off. He scrambled across the border to do some shopping for the only woman in his life.

"Dear mother," Dave wrote, "I went over into Mexico the other day and got you some silk stockings. They say they are the cheap kind. I will send them and see what you think of them." He also gave his mom "the latest dope" on his much-anticipated graduation, which was fast approaching. According to the Eagle Pass rules, parents were allowed to come down to the airfield up to five days in advance of graduation.

But Dave didn't advise it. "I will have a place for you to stay," he wrote, "but this is practically Mexico, so don't expect much. I hope the car is in good condition, for I am likely to want to get out of here in about nothing flat." That, at the very least, was Dave's plan. But word in the barracks was that the six guys in Dave's combat team, Team J of Squadron VI, had already been reclassified for their next assignments. Four of them were bound for single-engine school, one for twin-engine, and the other for instructor. Dave had no interest in being an instructor; few pilots did. But ultimately, the decision was not his to make.

"We kid a lot about becoming an instructor," Dave wrote his dad. "That is about the most useless position. It really gets some fellows hot under the collar when it comes to that. A lot of instructors use that to threaten the fellows. Ours does it a lot and it really gets some good laughs. I don't know what I would do if I get that. I am pretty dog gone young for a job like that."

When the CO learned that Dave's father was a lieutenant

colonel stationed at Camp Polk, Louisiana, he invited the chaplain to deliver the graduation address. Vaughn accepted. It was going to be a day to remember for the MacArthur family.

"Things are looking brighter than they have for a long time," Dave cheerfully recorded.

March 12, 1944

At 9:45 A.M., a large crowd spilled into the post theater. The feeling of patriotism, made into music by the Army Air Corps orchestra, was shared by all.

Parents flocked to Eagle Pass from all over the country, but none was more proud than Vaughn Hartley MacArthur, who identified his son among the two hundred graduates in Class 44-C. Even in a crowd of 630 people, Dave had an easy head of hair to spot—a genetic remnant of the MacArthurs' Scottish ancestry.

At Eagle Pass, graduation was a sacred ceremony, a tradition akin to the marriage ceremony. There would be oaths recited, pledges promised, tokens given, vows made. Pilots would receive silver wings and gold bars as a reminder of the duty to love, serve, and protect the nation till death did them part.

Just beneath the surface of the ceremony, on many parents' minds, was the fact that they may never see their sons again. The war was nearing its end, but no one at the time knew it.

In six days' time, the Eighth Air Force would deploy 679 bombers to destroy aircraft factories throughout Germany. Because the Flying Fortress had thirteen .50-caliber machine guns, it was widely believed that the aircraft could defend themselves against any enemy, especially when escorted by fighters. But forty-three of those bombers and thirteen escort planes would not return, courtesy of the Luftwaffe.

Eight weeks before Dave's graduation, General Dwight D.

Eisenhower became Supreme Allied Commander, Allied Expeditionary Force. A full-scale invasion of France, Operation
Overlord, was taking shape. By the end of the month, the Sixty-
seventh Tactical Reconnaissance Group would complete eighty-
three mapping missions over France, producing ninety-five
hundred photographs that would make possible the invasion of
Normandy.

The United States also landed in Italy, and under the command of Lieutenant General Ira C. Eaker, Operation Strangle
was now unfolding. The seven-week campaign would eventually deploy fifty thousand sorties and drop twenty-six thousand
tons of bombs on Italian railroads, ports, supply lines, and
other Axis-controlled infrastructure.

With body bags filling by the week, the demand for American pilots was at an all-time high. Every parent in the theater
knew it. The devastating news could be delivered at any moment. All it took was a telegram.

Henry Harley "Hap" Arnold, the General of the U.S. Army
and Commanding General of the Army Air Forces, made sure
every pilot knew the sacrifice his country asked of him. Later
regarded as "the General who invented the Air Force," Arnold
was one of the first military aviators in history and was taught
to fly by none other than Orville Wright in Dayton, Ohio. By
1944, some eighty thousand aircraft on six continents came
under his command.

In his foreword to *Wings over America*, which was standard
reading for every pilot, General Arnold wrote the following
charge:

> As members of the United States Armed Forces you
> do not have to be told of the magnitude and importance
> of the task that lies before you. At every base, station and

training field of the United States Army Air Forces you are preparing yourselves for the great test of arms which will prove that the forces of democracy can destroy the evil power of the totalitarian nations.

Soon you will take your place as Bombardiers, Navigators, Pilots, and Gunners alongside of our allies who have been fighting so valiantly. We can win this war, and we will win it, but only if every officer and enlisted man puts forth all the fortitude and resourcefulness that Americans have always displayed in times of war.

There are trying times ahead, times that will test the mettle of all of us, but I am confident that the personnel of my command will acquit themselves with honor and distinction, no matter where and when we shall meet the enemy.

At 10:40 A.M., Chaplain MacArthur took to the microphone. By the looks of him, you wouldn't know Dave was his son. There was not a hint of red in Vaughn's dark-brown hair, which he kept buzzed above the ears.

Vaughn stood a few inches shorter than his eldest son, and he was thicker in the chest than any of his boys. During the Great Depression, he had had a sharper jawline. But as his wallet filled out, so did his face.

The chaplain hadn't always been so clean cut. As a young teenager, he squandered his time as a hooligan, running wild with a gang of boys, getting into all sorts of trouble, and playing merciless tricks on the neighbors. His life changed one unexpected afternoon.

Chased through the countryside by an angry victim, Vaughn and his friends stumbled upon a large tent revival in an open field, typical of the time. They scurried inside and hid among the crowd. As Vaughn waited for the coast to clear, the words of the country preacher took deep root within him—words of freedom, peace, and sacrifice.

"A man cannot be a good soldier without character," Vaughn later wrote in a paper he submitted as part of his application into the U.S. Army chaplaincy. "Too much of our living has been selfish and self-centered. The army has no place for either." The goal, he wrote, "is to create in the mind of the soldier a confidence in himself and in his comrades." A soldier must never be bloodthirsty but a "friend of peace" who is "solely interested in maintaining the integrity of his nation and the protection of his people."

To his troops, Vaughn went by *Chaplain MacArthur*. To his fellow officers, he was *a good Joe*. But to Dave, who listened with childlike awe to his father's graduation speech, he was just *Dad*. And he was a hero.

Born in 1900, the elder MacArthur belonged to the Lost Generation that survived one world war and refused to retreat from the next. He began his storied service with the U.S. Army as a teenager, fighting the mustachioed Mexican revolutionary Pancho Villa in the third Battle of Ciudad Juárez on June 15 and 16, 1919. As a private first class, Vaughn was discharged from the Fifth Cavalry's Troop K three months later and went home to Massachusetts to marry Dorothy, his sweetheart. Dave had some big shoes to fill.

Chaplain MacArthur's address to Dave's graduating class was "Tinker to Evers to Chance," a title taken from a poem about a trio of professional baseball greats known for their successful double plays for the Chicago Cubs in the early 1900s: shortstop Joe Tinker, second baseman Johnny Evers, and first baseman Frank Chance. Listening to the chaplain's speech that morning was Eugene H. Butler, a fellow pilot in Dave's class who, decades later, still remembered the impact of the illustration. Vaughn "used them as an example of how success can be achieved when all work together for victory," he said. "He went on to tell us that if we didn't learn this invaluable attribute, we would not only fail, but be killed in our profession of Fighter Pilots."

On the dedication page of Dave's classbook, he and his fellow pilots penned their appreciation to all those who had shaped their silver wings, especially their instructors.

> *We are two hundred-odd men who undertook what has been the greatest challenge of our lives. We are not bragging when we say it wasn't easy, for not too many months ago playing tag in the clouds and ripping the sky wide open was but a dream. It was hard then to do more than dream. The real thing seemed a tough achievement to most; impossible to others. Finally, after hustling to keep a step ahead of Primary, Basic and Advanced— with heads still swimming—we have graduated.*
>
> *We're proud—"Pipe the neon on those wings!"—but we want to stop a minute and shake the hands of the guys who deserve the greater share of the credit. We snap our smartest highball to flight instructors, ground school and tactical officers, mechanics—everyone from the C.O. to the sarge who makes out the gig sheets.*
>
> *You took us to heart and taught us all you know. We won't let you down.*

The graduation ceremony concluded with a reception in the post theater lobby. Newly minted second lieutenants said their good-byes to their instructors, their friends, and their parents. Just where they would go after graduation was anyone's guess.

Some U.S. pilots traveled to the Far East, aiming their wings into Japan's Rising Sun. Others went north, patrolling the skies over Greenland. But for Dave, and also for his father, the shores of Western Europe awaited.

After sixteen months of rigorous flight training—from the palm trees of Florida to the scrub brush of Texas—Dave finally earned his wings. A wide smile spread across his face as his dad pinned the silver wings to his chest. A camera flashed. Congratulations were given.

The Eagle had finally hatched.

CHAPTER 6
THE CRACK-UP

June 16, 1944, Dover Army Airfield, Delaware

Dave would later call it a simple "crackup," but the mishap came with a $130,000 price tag.

He hand-turned the four-bladed propeller then climbed into the spacious P-47 cockpit. The "Jug" was among the heaviest single-engine fighters Dave could fly, weighing 17,500 pounds fully loaded. The United States produced more Juggernauts than any other fighter during World War II. By 1945, the Jug would tally 560,000 sorties.

The P-47 Thunderbolt was well-endowed with a supercharged turbine that fed oxygen directly into the engine. The complex air-intake system made for a quick takeoff, and at high altitudes it tricked the engine into behaving as it would at sea level. But the turbo wasn't immune to malfunctioning.

The preflight checks had gone off without a hitch. Master battery ON. Supercharger lever OFF. Fuel selector valve to MAIN. To get the juices warm and flowing, Dave primed the fuel boost pump, flipped the ignition switch, and after fifteen seconds engaged the starter. The engine awoke from its slumber, and the warbird sputtered to life.

Dave zigzagged to the line. He cracked the throttle a hair, set his turbo to FULL ON, and gave the ship the gun. It would be a flight he would never forget.

After graduating from Eagle Pass, Dave and his fellow squadron members were gung-ho for combat. But when the assignments had been divvied out, the young MacArthur was far from amused.

"Guess what?" he would later say during an interview, shaking his head with a smirk. "I was retained there at Eagle Pass as a P-40 instructor."

At the time, the news punched him in the gut. Instructing fledgling cadets was the kind of purgatory that even a Methodist had to believe in. To make matters worse, Dave hadn't logged a single minute of flight time in the plane he was tasked to teach, the Curtiss P-40 Warhawk.

In the Pacific skies, the P-40 Warhawk punched well above her weight class and developed a fierce reputation as the "Flying Tiger" in Japan. The P-40 had a rugged frame that could withstand heavy punishment. Pilots could strap on the airplane and perform violent aerobatics with a gravitational force that could crush the spine and tear the wings off weaker birds. The P-40's nose dipped low and resembled a jaw that many pilots painted with rows of razor-sharp teeth.

Much less intimidating, however, was the P-40 Dave was assigned to fly, a toothless trainer that answered to the name *Snafu 216*. Dave made a quick study of her, though, completing his checkup and instructing students on the basics of formation flying, gunnery, and bombing drills.

But combat was always on his mind. As often as he could, Dave went up to headquarters to request deployment. After six stubborn weeks, his incessant pleading paid off. "Darned if they didn't come up with an assignment," Dave would later say. He was finally sent to flight combat training in Dover, Delaware, and assigned to the First Air Force, First Fighter Command, 125th Army Air Force Base, Unit F.

For forty days, Dave would fly P-47s at the Dover Army Airfield. Once farmland, then a municipal airport, the airfield functioned as a Replacement Training Unit for the U.S. Army Air Corps. For pilots about to deploy for combat overseas, RTUs were the last hurdle.

Between the dusty hills of Texas and the coastal breeze of Delaware, Dave wrote his last will and testament.

I, David W. MacArthur, of the City of Boston, County of Suffolk, and State of Massachusetts, now in the military service of the United States of America, being of sound mind, make this my last will and testament.

All the rest and residue of my property and estate I devise and bequeath to my mother, Mrs. Dorothy B. MacArthur, 901 Dennis St., Leesville, La.

I hereby nominate Mrs. Dorothy B. MacArthur to be executrix of this my will and request that she be exempt from giving a surety on her official bond.

In the presence of three pilot buddies, Dave signed the document. He had little more than a wristwatch and typewriter to his name.

The torque of the P-47 was tremendous. Two thousand pounds of horsepower pushed Dave into the back of his seat as his wheels separated from the runway.

Stuffed into the nose of the plane was a Pratt & Whitney R-2800 Double Wasp engine. On a good day, it could produce top speeds of 433 miles per hour. On its best day, it could reach a whopping five hundred. But on its worst day, the "Flying Bathtub" could take a terrible tumble when its engine, a full ton of deadweight, sank the ship like an anchor to the ground.

The year before, to make room for seven incoming squadrons of P-47s, construction crews mounted an extensive upgrade to the Dover Army Airfield. They completed the airdrome's infrastructure, which previously amounted to little more than the foundations of a hangar, and they extended the main runway by pouring a canal of concrete some seven thousand feet in length.

When the crews encountered drainage problems on the airfield, they set out to dig a series of drainage ditches to alleviate the water. One of the ditches, a channel that emptied into a tributary and then into the nearby Little River, was dug particularly close to the runway. To keep the ditch from interfering with the planes, each section was to be covered. But unbeknownst to the workers, they missed one.

Shortly after takeoff, Dave knew he had a problem. His engine coughed a few times, gasped for breath, and then gave up its ghost. His turbo had run away.

The old adage, which every pilot knew, rushed into Dave's mind: If you lose an engine on takeoff, go straight ahead and put the plane on the ground. This emergency procedure was stated in no uncertain terms in the P-47 Thunderbolt Pilot's Flight Operating Manual:

Engine Failure During Take-Off

a. Nose down.

b. Landing on field STRAIGHT AHEAD. If too late, retract gear and land OFF field STRAIGHT AHEAD.

CAUTION
DO NOT ATTEMPT TO TURN BACK
INTO THE FIELD.

But directly in front of Dave was a fuel storage dump filled with explosive tanks. *I'm not going straight ahead into this!*

It was clearly a breach of protocol, but Dave had no choice. To reach the perpendicular runway on the airdrome behind him, he had to improvise. He knew that banking the Juggernaut would bleed off airspeed, but his windmilling propellers left him little choice.

When the slender strip came into view, Dave lowered his landing gear, locked his wheels, and pitched the nose forward, sacrificing altitude for airspeed. It was a costly trade-off. The plane dropped, picked up its best glide airspeed, and reached for its shadow, which grew larger on the ground. Dave glanced at the runway, which was still a good ways off in the distance. One hundred yards became eighty. Fifty became twenty. Dave pulled back on the stick, leveled his wings, and braced for impact.

The landing gear clipped first. Its wheels snatched the only uncovered section of drainage ditch. The Jug leaned into the punch, its nose connecting first with the ground. The force of the blow twisted the fuselage into a vertical spiral, tail in the air. The plane tumbled sideways, awkwardly turning over itself in a series of slow somersaults, one after the next, down the airdrome and across from the runway. After three horrifying cartwheels, the Jug lost her momentum and ground to a halt in the grass, upright and flat on her belly.

A dispatcher who was working in the operations center below the flight tower saw the crash and bolted from his post. He tore across the airfield on foot. If the pilot had survived, the poor chap would need all the help he could get.

Dave slid back the canopy. He was somehow unscathed. The worst of the crash was absorbed by the ⅜-inch-thick armor plates and 1½-inch-thick bullet-resistant glass designed to protect the pilot from gunfire. The turbine supercharger, with its elaborate ductwork inside the fuselage, also cushioned the cockpit. It was the very least the culprit turbo could do.

Dave slipped onto the wing and began filling out the damage report. The empennage behind him looked brand new, straight off the assembly line. Its rudder, elevators, and vertical stabilizer showed no signs of distress. The same could not be said about the underside of the bird. Its skin and talons had been torn clean off, exposing a mess of innards protruding from its side. The propeller was now shaped like a swastika, its blades bent sideways like rheumatic fingertips forced out of joint. The mechanic crews would definitely have their work cut out for them.

The dispatcher arrived at the site of the crash, exhausted from his heroic sprint and gulping for air. He doubled over on the field, hands on his knees. Seconds later, the ambulance and fire truck pulled up. The crew shot from their rig and looked frantically for the pilot. They saw the wheezing dispatcher and pulled him onto the stretcher. From his perch on the wing, Dave watched the scene unfold and laughed like hell.

"I'm not the pilot," the dispatcher blurted, finally catching his breath. He pointed to the disfigured P-47. "He's sitting on the wing over there."

The ambulance crew took one look at Dave and drove away.

Dave's crack-up had a happy ending, but the same couldn't be said of every mishap that occurred during flight combat training at the Dover Army Airfield. During a three-and-a-half-year span, roughly 180 P-47s would take similar spills. Some of the crashes did little more than sully the flyboys' pride. Others, however, were fatal, sobering reminders of the price that pilots had to be willing to pay.

For Dave, this would be his first, but certainly not his last, odd mishap in a plane. Years later, long after the world was again at peace, another of Dave's landings would send witnesses barreling toward his runway in panic. On that day, Dave was flying his beloved T-6 along the New Hampshire seacoast when an air projectile slammed into the Plexiglas windshield of his canopy. The missile—if one could call a duck that—exploded in a burst of feathers, knocking off Dave's helmet and blinding him with blood. Somehow, he again managed to land his plane. When the first attendant arrived at the scene, he vomited, not knowing whether the bony goo that dripped from the helmet belonged to Dave or the unfortunate fowl.

In May, Dave logged seven hours and thirty-five minutes of flight time in the Juggernaut. By the end of June, after flying almost every day that month, he added nearly fifty-four more hours.

On July 7, the Army Air Forces deemed him ready. Following a brief leave back home in Leesville, Dave's parents dropped him at the train station in New Orleans and waved good-bye to their firstborn son. Dave transferred to the Richmond Army Air Field in Virginia for deployment overseas in a combat unit. The newly appointed second lieutenant was only days away from boarding a troopship for occupied Europe. After a final letter to his parents, he readied himself for war.

Dear Folks,

Just to let you know that I'm in Richmond. Am reporting out to the base in about an hour. It should not be more than five to ten days now. I will let you know if possible where I am. I doubt if I will be able to tell you. I will telephone you, if possible, when I leave here. Just as soon as I get my portable typewriter again, I will write a letter.

If you can possibly get a camera in the next day or two, I wish you would get it and send it to me, but quick. I am not hard to please now. Will write more later.

Love, Dave

CHAPTER 7
BLITZKRIEG

Dave arrived in the Bay of Naples in late July 1944. Mount Vesuvius, towering above its waters, turned a dazzling emerald in the sunlight. Like a sentinel, its eye guarded the vineyard-laced landscape, watching the Allies pour onto its coast. Every century or so, when humanity grew too big for its britches, the volcano knew how to humble it.

During the famous eruption in AD 79, Vesuvius vomited a column of smoke that rose twenty-one miles into the atmosphere—more than twice as tall as the atomic clouds that would soon mushroom over Nagasaki and Hiroshima. When the cloud collapsed, a tidal wave of molten rock, ash, and gas surged down the volcanic slopes at 430 miles per hour, as fast as Dave's P-47 Thunderbolt. Anyone caught in its path, including the doomed citizens of Pompeii and Herculaneum, were enveloped by the 1,830-degree inferno, their dying facial expressions preserved for future generations.

Only six days after Dave graduated from Eagle Pass, Vesuvius erupted again. A B-24 tail gunner flying near its plume caught the phenomenon on film. At the time, the 340th Bombardment Group of the U.S. Army Air Forces was stationed at the Pompeii Airfield. The volcano showed no mercy to the base, mortaring an artillery of cinders, rock, and smoking debris. The engines, surface controls, windscreens, and gun turrets of some eighty-eight B-25s were damaged.

Some intellectuals speculated that the Allies had triggered the eruption by bombing the unstable earth surrounding Mount Vesuvius. The idea was "highly problematic" for one geology professor at Colgate University in New York "but well within realms of possibility." There was even talk of triggering similar eruptions in Japan, possibly kindling Mount Fuji.

In the meantime, Dave had been assigned to the Fifteenth Air Force, established in November 1943 under the command of General Jimmy Doolittle. The day after its founding, pilots lifted off from air bases throughout southern Italy, taking the fight directly to the Germans. Dave was accustomed to flying—and, at times, crashing—the heavy P-47. But in his new combat assignment, the outfit he was assigned to didn't have P-47s. Instead, they flew P-38s.

In the four months leading up to Dave's arrival in Naples, his new unit flew an enormous mission over Ploesti, Romania, whose oil refineries produced roughly a third of the Axis's oil. The mission—Operation Tidal Wave—obliterated the refineries, but the United States suffered its greatest loss to date of aircraft and airmen in the effort. Soon Ploesti ramped up production again, forcing the Allies to continue bombing the city.

Although by the end of the war those efforts would prove successful, in the spring and summer of 1944 Dave's new outfit was nearly destroyed in their stab at Ploesti. From April until July, the fighters accompanied bombers on their raids in P-38 Lightnings. They went in as dive-bombers and, over the months-long mission, were nearly wiped out. What was left of the unit went to North Africa to reequip.

When he arrived in Naples, Dave was flown to Africa as well, and he stayed with the outfit as they got reequipped with P-38 Lightnings. While there, instead of flying P-47s, Dave got the chance to fly a plane of equal beauty, the legendary Supermarine Spitfire.

The slim-and-trim Spitfire was a racehorse bred for speed. Her elongated nose, stressed aluminum skin, and featherweight frame was pulled through the air by a 1,000-hp engine. "It just didn't seem to want to slow down," one Spitfire pilot said. "When one pulled back on the throttle, it took a long time to take effect on its speed."

The unique shape of her wings, elegant and elliptical, gave the Spitfire the high-altitude athleticism she needed to spar with Germany's Focke-Wulf 190 and the highly decorated Bayerische Flugzeugwerke 109.

Four years earlier, in May and June of 1940, fifteen Spitfire squadrons had provided air support over Dunkirk. After the evacuation, the iconic plane captured the popular imagination and became a national hero, a symbol of strength and resilience that represented the best of the British spirit. The Spitfire furthered her credentials by assisting the Allied air forces in the months leading up to D-Day. By April, the sky was crowded with British Spitfires and American P-51 Mustangs. For every one aircraft flown by the Luftwaffe, there were thirty Allied to greet it.

By June 1944, the invasion of Normandy was the world's best-kept secret. To prevent the Germans from telegraphing the attack, inflatable tanks were positioned in Kent, directly across from Pas de Calais, the shortest distance between the English and French coasts. Nazi reconnaissance planes reported no unusual activity over the English Channel. The water was choppy. The skies, capricious. The tide wasn't in favor of an amphibious landing.

The beaches along the French coast remained virtually virgin of bombs until the very last minute. Then the Allies would surprise the enemy with overwhelming force. The war tactic was taken out of Hitler's own playbook, a strategy the Führer called *Blitzkrieg*, or "lightning war."

All eyes, especially those belonging to the pilots, were on the weather. In a memorandum on June 3, General Dwight D. Eisenhower wrote, "Success or failure might easily hinge upon the effectiveness, for example, of airborne operations." The skies had to cooperate.

On June 6, 1944, when D-Day dawned, 862 heavy bombers flew across the English Channel, dropping their payloads over enemy fortifications in France. Low-ceilinged stratocumulus clouds covered the coast, but the Allies brought their own weather with them—175 P-47 Thunderbolts and 46 P-38 Lightnings.

By H-hour, 6:30 A.M., German gunnery kicked into gear. They aimed at the waterline, belching out their bullets at the shore-bound soldiers. A P-38 pilot flying above Omaha Beach looked down and saw the carnage. "German resistance appeared to be devastating," he later remembered. "Landing craft were being capsized, some were exploding, and the contents, including men and equipment, were being spilled into the surf in great numbers and quantity."

But the skies belonged to the Allies. "If you see fighting aircraft over you," Eisenhower assured his troops, "they will be ours." The future U.S. president was right. Above the land invasion, or what one major general described as the "historic epicenter" of the U.S. Army, a grand total of 3,467 heavy bombers, 1,645 medium bombers, and 5,409 fighters streaked across the sky.

Not a single plane was lost.

After Dave stretched his Spitfire's wings, the Fifteenth Air Force awarded him the choice of a lifetime: He could fly either the P-51 Mustang or the P-38 Lightning. It was hardly a choice at all.

"I always wanted to fly P-38s," he later said. "That was the prize." The Lightning first caught Dave's eye during basic training in Coffeyville, Kansas. "But only the hot ones get that," he had written his dad.

On August 19, 1944, after returning from North Africa, David MacArthur became one of the hot ones. He was assigned to the Fourteenth Fighter Group, the same group that had been responsible for patrolling the coast of California after Japan's attack on Pearl Harbor. As part of the group's Forty-ninth Fighter Squadron, Dave was sent to the spur of Italy's "boot" and stationed at Satellite #7 of Triolo Airfield. About seven miles away was San Severo, a charming little coastal city founded in the eleventh century, roughly a thousand years after Vesuvius wiped Pompeii off the map.

The Triolo Airfield was part of the Foggia Airfield Complex seized from the Luftwaffe in 1943 and repaired by the U.S. Army Corps of Engineers. It was a strategic nerve center for U.S. and British forces overseeing the movement of ground troops in Italy, naval squadrons in the Adriatic Sea, and U.S. pilots contesting the Mediterranean skies.

When he arrived at Triolo, Dave's knowledge of the P-38 was vague at best. His buddies had flown Lightnings in Africa, but he'd been in Spitfires. Dave was given a handbook and told to sit in the cockpit and learn the plane. Nobody was around to answer questions except his crew chief, and he didn't know anything about the flight characteristics. So Dave sat, and he learned.

Soon he was slated for a one-hour "piggyback" familiarization ride in the Lightning. He was officially listed as copilot in his flight log, but in reality Dave was only a passenger and a cramped one at that. The P-38 was definitely a one-pilot plane. When the radio was removed from behind the pilot's seat, there was just enough room for a man if he curled up in the fetal position.

With his knees pulled to his chin, and without so much as a finger on the controls, Dave squeezed into a ball and entered the bubble canopy for the most claustrophobic ride of his life. The parachute on his back made the space feel smaller, but since there wasn't a safety strap holding him down, Dave was glad to be wearing it.

Sixty seconds as a "pig" felt like a full century of bumps and bruises. The plane seemed to do everything in its power to re-sist reuniting with the runway. Without safety straps holding him to the seat, Dave was jostled this way and that. When at last he landed, Dave was welcomed with applause into the fra-ternity of the "Unholy Order of the Piggy-Back." The rite of passage even came with a membership card, which confirmed that the "piggy-backer" was "of unsound mind and question-able sanity."

It was an awkward experience never to be repeated, but the bird was such a beauty. Lockheed's twin-boom design had taken the world, and the war, by surprise. The P-38 shattered Howard Hughes's record-breaking speed flight from Los An-geles to Newark, a distance of 2,490 miles, soaring 100 miles per hour faster than any engine in the sky. It was the first fighter to top four hundred miles per hour in level flight, and in a dive, the plane approached the speed of sound. Unmatched in stability, the Lightning allowed photo-reconnaissance pilots to capture the shots that allowed detailed maps previously thought impossible. About 90 percent of all aerial photos taken over Europe in World War II were snapped by the swift P-38. In only six weeks, the "Photo Joes" of the Twelfth Photo Reconnaissance Squadron of the Third Photographic Group mapped up to 80 percent of Italy.

The P-38 Lightning was the first plane to shoot down an enemy aircraft after the United States entered the Second World War. In his radio address on September 11, 1941, President Roosevelt gave permission to the U.S. Navy to "shoot on sight" any German U-boats spotted in the west Atlantic waters. Almost one year later, on the morning of August 14, 1942, Second Lieutenant Elza Shahan took to the skies above Iceland in his P-38.

As part of the Twenty-seventh Fighter Squadron of the First Fighter Group, Shahan's mission was to intercept a Focke-Wulf Fw 200 Kurier that was up to no good in the area. The four-engine German bomber was flying reconnaissance for the Luftwaffe, relaying U.S. activities to nearby U-boats. Shahan came up behind the Kurier, aligned his gunsight, and released his 20mm cannon. The rounds connected. An engine exploded. The Lightning had drawn first blood.

After his makeshift piggyback ride in the cramped P-38 cockpit, and only twelve days before his twentieth birthday, Dave was on the hunt for his own kills, soaring solo across Italy in the wings of his dreams. The bases of the Fifteenth Air Force stretched all the way across the country, an intricate web of Allied airfields synchronized for missions throughout the European continent.

One warm day in mid-August, a crippled P-47 Thunderbolt landed at the Triolo Airfield, desperate for repair. The mechanics patched her up and needed a pilot to give her a test flight. Since Dave had trained in the Jug, and knew the plane inside and out, he was tapped as the man for the job. Dave lifted off the runway, stretching the Thunderbolt into the horizon, unaware of the trouble waiting for him in the distance. Years later, Dave would give an innocuous description of the perilous test flight. "I got in a little dogfight thing," he said.

An hour after Dave arrived back at the base, his squadron commander called him up to his office.

"What'd you do on that flight?" he demanded.

"Nothing," a sheepish Dave replied, knowing full well his "little dogfight" had some unintended consequences.

"Well, somebody buzzing in a P-47 just took down the whole antenna system for the Fifteenth Air Force, and they think it's you." The CO rubbed the back of his neck and furrowed his brow. "I'm giving you a jeep," he said, "so get your butt out of town and don't come back until I tell you to."

Dave threw some clothes into a bag and took off in his jeep for Naples, some 120 miles away. He was to call the CO every other day.

With an unexpected break, and never one to pass on a good time, Dave made his way to the Orange Grove, the Allied Officers' Club perched on the edge of a cliff overlooking the Bay of Naples. U.S. troops had commandeered the famed Giardino degli Aranci, a ritzy private club that now hosted all stripes of officers for dancing and drinks each night of the week. It was a welcome respite for the war-weary Allies.

A lush garden of orange trees and colorful fountains filled the open-air marbled terrace as the club's Italian orchestra tried its hand at American swing. In the distance, across the Bay of Naples, sat Mount Vesuvius, still simmering from its recent purge. Dave traded a dollar for a membership card, made for the bar, and struck up a conversation with a fellow Bostonian, a Catholic named Murphy.

Two months earlier, on June 4 and 5, Mussolini's Rome had fallen into the hands of the Allied forces, spearheaded by the Fifth Army. It was the first of the three global Axis capitals to crumble. After the liberation, Pope Pius XII held an audience for the Catholic Allied troops who curved around the Coliseum in their tanks, crossed the Tiber River, and swept into St. Peter's Square. The Vatican maintained official neutrality throughout the war, angering many Allies around the world. But despite his public silence, the fifty-eight-year-old Italian pope, born Eugenio Maria Giuseppe Giovanni Pacelli, worked behind the scenes to aid the victims of Axis atrocities and lobby on behalf of peace while staying in close contact with President Roosevelt, members of the German Resistance, and Winston Churchill. In doing so, he had not escaped the ire of Adolf Hitler.

On July 26, two hours after Hitler received word of Mussolini's demise, the infuriated Führer suspected treason and blamed the pope for the fall of Rome. "I'll go right into the Vatican," Hitler told Walther Hewel, liaison officer to the Nazis' foreign minister. "Do you think the Vatican embarrasses me? We'll take that over right away. . . . We'll get that bunch of swine out of there. . . . Later we can make apologies."

But Germany was forfeiting territory in Europe. Now, in August, two months after Rome's liberation, Churchill traveled through Italy to survey the continued Allied progress in the Mediterranean. Paris was also on the brink of liberation. The French Resistance mounted their uprising against retreating German forces as General Patton's Third Army crossed the Seine and approached the city. There was much to celebrate.

"Let's go to Rome!" Murphy erupted. He had missed one chance to see his pope and wasn't about to miss a second.

Dave wasn't Catholic, but he was the one with the jeep, so the two men drove up the Italian coast and made their way to the Eternal City. They "liberated" a room at the swanky Regina Hotel in Rome and settled in for the night. When morning came, Murphy quickly donned his uniform. It would be a monumental day at the Vatican, a day that even a Methodist could appreciate. Dave decided he might as well tag along.

A thronging multitude of people flooded into St. Peter's Square, spilling around the two fountains and the Egyptian obelisk in the center. Statues of martyrs, standing tall above the colonnade, gazed down in sympathy at the troops crammed within the circular plaza.

Dave and Murphy muscled through the crowd, inching toward the basilica to get closer to the action. If they were going to have to endure the pomp and pageantry, Dave reasoned, they might as well get good seats. Swiss guards holding medieval pikes flanked the portico. Their uniforms, a flamboyant mix of red, purple, gold, and blue, looked entirely untouched by time, something straight out of the Renaissance.

"We've come all the way over from Foggia," Dave explained to one of the Swiss guards. Inexplicably, the guard nodded and motioned for the two men to follow. "Come with us."

The delighted Americans trailed their escort up the basilica steps, over the threshold, and down the central nave that stretched like a runway all the way to the angel-crested altar beneath the dome. Dave took his seat with the cardinals at the back of the papal platform. Beside him was Murphy, who by this time was beside himself with euphoria.

A hush settled over the sanctuary, followed by faint shuffling noises that grew louder and louder. Dave craned his neck to find the source of the sound. Twelve slippered footmen, dressed in red robes, carried the ceremonial throne on which Pope Pius

XII sat. With a white shoulder cape covering his watered-silk cassock and a white cap hiding his thinning hair, his Holiness smiled from behind round eyeglasses and blessed the congregation as his footmen shouldered him toward the platform.

In seven different languages, Pius XII spoke to the gathering. When he finished, the Swiss guards cordoned off the platform and waited for the audience members to leave. In the meantime, the pope circulated among the dignitaries and eventually turned to greet the troops on the platform. The uniforms Dave and Murphy were wearing caught his eye. "You're Americans!" he said, walking in their direction. The pope offered his hand and Murphy knelt to kiss his ring. He then faced Dave and again extended his hand. Dave grabbed it and shook it heartily.

"Did you bring a rosary?" the pope asked.

"No sir," Dave said. "I'm Protestant. But I enjoyed your speech very much!"

Pius XII smiled, reached into his robe, removed two Vatican coins, blessed them, and handed them to the Americans. Then he motioned to a Swiss guard, asking him to take the young men on a tour of the Vatican and deposit them at his papal chambers later in the day. The two boys from Boston would have a private audience with the pope.

The behind-the-scenes tour lasted hours, a blur of botanical gardens, ancient artifacts, and priceless treasures stockpiled over the centuries. They breezed by hallways decorated with painted maps and hanging tapestry, by rooms filled with semi-clad human sculptures with amputated limbs and washed-away faces. Then they climbed a flight of stairs and entered the Sistine Chapel. Above them stretched a ceiling made celestial by Michelangelo four hundred years before.

After the trek, Dave and Murphy were escorted to the Apostolic Palace, only moments away from their private audience with Pius XII. But when they reached the papal chambers, the plans quickly changed. The pope was occupied by another guest who had traveled all the way from London. His name was Winston Churchill.

Dave and Murphy left Rome and again drove south, bypassing Naples by a few miles to see the archeological ruins of Pompeii. They strolled the eerie streets, weaving through the hollowed-out houses. In 1912, in the house once belonging to a priest named Amandus, excavators had made a significant discovery—a blue fresco painted onto a plastered wall.

The fresco depicted Icarus, the Greek mythological aviator who escaped captivity in Crete by flying over its water with wings given to him by his father. As the legend goes, Daedalus constructed the wings using feathers and wax, and when Icarus flew too close to the sun, his wings melted off and Icarus crashed into the Mediterranean Sea, where he drowned.

Dave left Pompeii, dropped Murphy back in Naples, and phoned his CO. The communications system of the Fifteenth Air Force had been restored, and it was safe to return to Triolo. Dave was no longer in the crosshairs of the brass. Three weeks later, and with fewer than twelve hours of flight time logged in the P-38 Lightning, Dave was tapped to escort bombers in his first combat mission of World War II. On October 6, on his fifth mission, he would find himself flying over the Adriatic Sea, a mere five hundred miles from Crete.

CHAPTER 8
THUNDERING HERD

Camp Polk, Leesville, Louisiana

If Vaughn MacArthur read the local newspaper on the morning of October 14, 1944, he would have seen a flurry of headlines. The Allies had pushed the Reich deep into Belgian territory. Americans were fighting house-to-house in Germany, closing the gap in the Battle of Aachen. In Paris, which had been liberated on August 25, the first American film since 1940 was shown, the romantic musical comedy *It Started with Eve*, starring a nineteen-year-old Deanna Durbin. More sobering, thirty-six B-17 Flying Fortresses and twelve fighter escort planes from the U.S.'s Fifteenth Air Force were reported missing.

Chaplain MacArthur sat at his typewriter in his Leesville home, plucking out an overdue letter to his eldest son. The family dog sat on the low sofa beside him, nipping at his hand. Outside, the temperature had reached a pleasant seventy-three degrees, warmer than it had been but not unusual for autumn in central Louisiana.

The very fact that Vaughn was in Leesville at all was rather remarkable. A fortuitous series of events, unfolding at just the right time, brought him to the camp nestled in the sultry Louisiana swamps. Three years earlier, in 1941, Chaplain (Capt.) MacArthur was assigned to the Forty-sixth Armored Infantry Regiment of the Fifth Armored Division. He had completed nearly one year of extended active duty in the U.S. Army's Chaplain Corps and, on the cusp of a possible promotion to major, had a difficult decision to make—remain in the military or return to civilian life. For starters, Vaughn had exceeded his shelf life. When it came to chaplaincy appointments in the Regular Army, new regulations placed a premium on younger men. At forty-one, Vaughn was past his prime.

The turning point came on September 11, 1941. With three sons soon to put through college and having taken a pay cut when he left his church in Buffalo to rejoin the Army, Vaughn was in desperate need of advice. He typed a letter to Chaplain (Maj. Gen.) William R. Arnold, Chief of Chaplains of the U.S. Army Chaplain Corps.

Subject: Reappointment and Promotion
To: Chief of Chaplains

1. On March 18, 1942, my commission as a Captain in the Chaplain's Corps expires. On that date, my time in grade is up which together with my certificate of capacity (granted October 1940) makes me eligible for my Majority.
2. On March 13, 1942 I will have completed a year of extended active duty. I have advice from your office that because of my age (41 yrs.) there is no chance for me in the Regular Army Chaplaincy.
3. I have a wife, and three boys to put through college in

the next few years. A Majority now would help compensate the loss of about $200–300 a year salary and honorariums suffered when I left my church in Buffalo, N.Y., and at the same time help me meet the college expenses of the next seven years.

4. The Bishop, Charles W. Flint, is willing that I remain in the service as long as I am needed but I cannot expect a church to accept me as its Pastor when there is danger of my being called into active service.

5. Being 41 yrs. old naturally conflicts with the recent desire for younger men in the Chaplaincy, and at the same time jeopardizes my future promotions in our church because like the Army, the church needs younger men.

6. In the light of the above five facts, I am asking for your personal advice as it relates to my future in the Army and in the church.

(a) Is there any certainty of a promotion for me in the next month when I shall have completed the requirements?

(b) Is there any chance that a man of my age might get a Regular Army commission in the Chaplain's Corps?

(c) Would you advise me to accept reappointment in view of the fact that my year is completed next March?

(d) Will I be held over my year of extended duty?

7. Chaplain William D. Cleary, our Armed Forces Chaplain, expects to be in Washington, D.C., before the month is over. He knows something of my work. I shall await his return before making my final decision on this most important matter.

Vaughn H. MacArthur
Chaplain (Capt.) 46th Inf. (Armd.)

A response to Vaughn's letter came a week later, but it offered little clarity.

Dear Chaplain MacArthur:

We have your letter of September 11, 1941, and we pro-
ceed to answer your questions in the order in which they
appear in this correspondence.

You will be eligible for promotion upon the recommen-
dation of your commanding officer and if there are any
existing vacancies in the allotment of officers in your
unit at such time. Only those clergymen who have
reached their 23rd birthday and have not passed their
34th birthday are eligible for appointment in the Regular
Army Chaplaincy. Whether you will be retained for an
additional tour of duty at the expiration of your present
term is difficult to say. Only conditions at that time will
determine your future status. If you desire to be retained
on active duty you should request your commanding officer to initiate action to extend you for an additional period of one year. He in turn will forward it to the
commanding general of the area in which you are serving, who will recommend to The Adjutant General that
you be extended or issued the necessary orders extending
your tour.

For the present the decision to remain on duty is a matter
for your own judgment. Although there is a need for
chaplains, it would be very unwise for this office to tell
you what to do in this important matter.

For and in the absence of the Chief of Chaplains:
Carl L. Wilberding, Chaplain Assistant

Within a matter of months, the decision had been made. Vaughn was promoted to major, retained for an additional tour of duty, and assigned as chaplain of the newly activated Fifth Armored Division of the U.S. Army. By early 1942, he moved with his division to Camp Cooke, California.

On May 8, 1942, Vaughn was relieved from duty at Camp Cooke and ordered to Fort Knox, Kentucky, the new home of the fledgling Eighth Armored Division tasked with guarding the vaults of the United States Bullion Depository. The impregnable fortress of granite and concrete safeguarded the nation's most precious treasures—the U.S. gold reserves, the Declaration of Independence, the U.S. Articles of Confederation, an early copy of the U.S. Constitution, and Abraham Lincoln's 1863 Gettysburg address. Fort Knox was also home to the burgeoning industry of armored warfare.

Though Winston Churchill had sworn under oath in 1925 that H. G. Wells was the first person to dream up the idea of an armored vehicle—what the sci-fi writer coined *the Land Ironclad*—Leonardo da Vinci actually beat him to the punch some four hundred years earlier. In 1487, the prescient Italian genius sketched what he called *the covered chariot*. Though it never materialized, his blueprint called for a steel-plated, four-wheeled flying saucer–looking machine powered by eight men on the inside, who could shoot 360 degrees of fire from guns mounted around its periphery.

By World War I, technology finally caught up with the European imagination. Trench warfare, which often ended in stalemate, created a problem on the battlefield that engine-driven "land ships" promised to solve. Though riddled with mechanical defects, the rudimentary British Mark I, first used in combat against the Germans at the Somme in 1916, was a behemoth to behold. Its twenty-six-foot-long caterpillar tracks could roll right over trenches and penetrate deep into enemy lines. The "tank," so named for its resemblance to steel water tanks, became the new cavalry. The day of the horse was over.

In the 1930s, with a modest military budget, the United States lagged behind the British, French, and Germans in military might. In contrast to the swelling German army, which by the end of the decade was a half-million strong, the United States paled in comparison, totaling 190,000, half of whom were deployed overseas.

War was again looming on the horizon. President Roosevelt enacted the Protective Mobilization Plan on September 8, 1939, which was created, he said, "for the purpose of strengthening our national defense within the limits of peacetime authorizations." The United States needed more tanks and more troops. Soon, assembly lines were producing tanks en masse, including the mainstay M4 Sherman. But training men for armored combat was a more challenging endeavor. In World War I, the United States learned that tanks perform best if supported by infantry soldiers. This realization demanded new war tactics and training programs.

Fort Knox spearheaded this training, and on July 15, 1940, the War Department established the First Armored Division at the Army post. Vaughn's first Army orders sent him as a chaplain to this newly formed division, and two years later, his orders deposited him back at Fort Knox to join the Eighth Armored Division, *the Thundering Herd.*

Activated on April 1, 1942, the Eighth Armored Division spent nine months training cadres of personnel. They were the first official division to guard the gold vaults at Fort Knox before relocating to Camp Campbell, Kentucky, and then to Camp Polk, Louisiana, which by the end of the war would train some eight million men.

Two years earlier, the youthful Colonel Dwight D. Eisenhower had never seen combat, but his frustration with the "snail-like progress" of World War I tanks brought him to central Louisiana to assist in choosing the site for what would become Camp Polk. Riding on horseback seven miles east of Leesville, Eisenhower dismounted, planted his walking stick into the soil, and declared, "This is the location where the new camp will be built!"

The farmlands, once known only as Hog Wallow, perfectly simulated the soggy fields of Europe. In the rain they became "terrifically muddy," ideal terrain for troops learning how to counter Hitler's *Blitzkrieg*. But no one could have predicted the central role the camp would play in the ultimate defeat of Nazi Germany, especially not the Leesville farmers ducking beneath the occasional stray bullet.

A year after his prophetic declaration, Eisenhower returned to Camp Polk, along with his good friend George Patton, to participate in the Louisiana Maneuvers. It was the largest war games ever simulated in the United States and would soon be recognized as the most effective method of preparing U.S. soldiers for European combat.

By this time, Patton was a man of sixty, ready to retire. By all accounts he had reached the end of his military career. But the veteran threw himself into the training exercises, drawing from experiences he had gained as the commander of the first U.S. tank brigade in World War I.

In September, in the most complex and important of the Louisiana Maneuvers, a "Red" army of two hundred thousand men fought a "Blue" army of equal numbers across a battleground that stretched from Shreveport to Leesville, some 3,400 square miles. After five grueling days of combat, *Time* magazine reported on the outcome: "So ended the first phase of the Battle of Louisiana."

The famous Louisiana Maneuvers gave Patton renewed energy. Both he and Eisenhower went on to replicate in Europe what they had practiced in the steamy swamps of central Louisiana. Eisenhower would carve his path to the U.S. presidency. Patton, never one to endure a desk job, would be solidified as one of the greatest military leaders in U.S. history.

Vaughn MacArthur was whittled from the same Louisiana wood.

By September 1944, the Eighth Armored Division was packing up for war. They had been prioritized for overseas deployment and placed under the capable command of Brigadier General John M. Devine. "I don't think I've seen a better-looking group of soldiers who were better trained than this division," the general said.

The official alert for transition to the port of embarkation came in October. With their training complete, they crated their equipment, and the sixteen units of the Eighth Armored Division—262 officers and 2,558 enlisted men—scrambled to put their affairs in order. It was only a matter of days.

The atmosphere at Camp Polk on October 14, 1944, swirled with anticipation. As Chaplain MacArthur sat down to type his son a letter, a storm was brewing far to his south as well. In the skies above Nicaragua, strengthened by the warm waters of the Gulf, the cyclone would soon mature into a Category 4. History would name it "The 1944 Cuba–Florida Hurricane." The path of the hurricane wreaked havoc on many of Dave's old training grounds. It rotated like a propeller over Miami, up the eastern coast to Kitty Hawk and Richmond, before banking east into the Atlantic. The storm followed his son to sea, and in a matter of weeks, Vaughn would follow the storm.

Leesville La
14 Oct 1944

Dear Dave,

I feel like a heel, for this is the first letter which I have
written or had written this week, and there hasn't been
any reason for it only for our present state of rushing
around getting all the equipment which we need for the
winter. I have been hunting around for a combat suit, for
they tell us that European weather is like Louisiana
weather and they are good in either place. They are not
issued but we have to chisel them, so I spent quite a few
hours trying to chisel.

To add to the confusion, we have just lost a chaplain,
and this afternoon as I was ready to come home I
received the new one. This fellow is a Catholic Chaplain
and comes from Indiana. He is fairly green, so all our
Catholic Chaplains will learn the hard cold way. I am
trying to get all the chaplains outfitted and equipped with
all the stuff that is necessary for their work.

I hope you have been reading between the lines, because in that case you will not be surprised when you get
our change of address. Mother has been down here too
long and, as a result, Charlie's education program has
suffered, so we have come to the place where he has to
get to a better school if he is going to get into college. I
shall certainly miss them, but they have to go some time
sooner or later.

We all will be miles closer when they head for Boston.
Grammy will want them to move in with her, but I don't
think it healthy, neither do I think it wise for them to go
in with Aunt Bernice. I want both of them to get their
schooling under way and have it all over by the time we

return from Europe so that your schooling will be picked up on your return and we can help you in it.

Today I put in a claim for the camera and the flash gun and the films which I included in the package when I mailed it to you at Richmond. I am not sure that I put the right address on it, and that may have been the cause of the delay in receiving them. I put the 125th rather than the 120th, but the report said it was shipped to the APO 16304. From there they were to ship it overseas, but it is evident that somebody hooked onto it, and so I hope that the insurance holds good, for it cost me $50 bucks.

Charlie is going to see if he can pick up a camera for you in Boston and then he can ship it over to you. I am carrying 24 rolls of 35 mm just in case they ship the thing to me overseas. In case they ship you this one and Charlie gets you one, you can always sell the one you do not wish to keep.

We have been shipping a lot of stuff over to you, but as we cannot insure the stuff, you cannot be sure that it will get over. I understand that stuff is being stolen right and left, both G.I. and civilian stuff.

I guess I told you before that we got your typewriter repaired, and they did a good job on it. That will mean that you will have to hurry back so that you can use it. We are lucky, for they have issued us two typewriters, and we have them all fixed for the trip. I ought to be able to do some writing when we get over, but I understand that the air mail envelopes are rationed to us.

This afternoon we have been taking the pictures that we have taken since we have been in the Army, and we have been naming them and the places. It is funny how easy it is to forget some of the places and some of the names.

Mother is packing some of the surplus stuff that we have, and we have been going over some of your stuff this afternoon—this blooming dog is on the divan beside me biting at my hand as I try to type. He has a sore on his back, and I don't know what it is caused by. Mother will have to take him out to the Vet tomorrow.

. . . We have a pretty good supply of magazines, for in the time that I have been meeting trains, I have given away better than 50,000 magazines. Beside that, they are giving away games on the trip so that they will be occupied and kept out of mischief.

Now that I have put in a claim for the camera, I expect that you will be receiving it, and I hope that is the case because they are so hard to get these days. I am going to get some leather gloves for you, and also send you a field jacket that I have. That is about the only thing that I can think of to send you for Christmas, and tomorrow is the last day in which we can send you stuff without your asking for it.

Well I shall try to do better. There will be a break in the letters soon while Mother is visiting Boston, but I hope that she will not forget us and keep writing to us whether she hears from us or not. If we ever go over, you can see by the address where we shall be.

> I'll be seeing you.
> Love, Dad

Vaughn wouldn't get the chance to send his letter.

A young man rode his bicycle through the quaint neighborhood in Leesville. When he arrived at his destination, 901 Dennis Street, he walked through the yard, up the front path, and knocked.

A short woman with curly brown hair and round glasses opened the door. Her eyes flew to the Western Union telegram in the boy's outstretched hand and her stomach sank.

THE SECRETARY OF WAR DESIRES ME TO
EXPRESS HIS DEEP REGRETS THAT YOUR
SON 2ND LT DAVID W MACARTHUR HAS BEEN
REPORTED MISSING IN ACTION SINCE 6
OCTOBER OVER GREECE. IF FURTHER DETAILS
OR OTHER INFORMATION ARE RECEIVED YOU
WILL BE PROMPTLY NOTIFIED.

PART II

To each man in combat there is some experience that will remain with him for the rest of his life, and I had mine on the sixth of October in 1944. That was the start of an experience which lasted for over eight months, and which I will never forget. It is interesting enough, I think, to warrant a few words. Actually it was a minor mission into Greece, but it happens that minor missions, or "milk runs," as they are called, can turn into some of the biggest missions in the books. This one turned out that way for me.

—DAVID W. MACARTHUR,
1946

CHAPTER 9
THE FORK-TAILED DEVIL

One week earlier, October 6, 1944

"Mac, pull up. You're on fire."

A lucky round of 44mm German antiaircraft shells punctured the thin skin of Dave's P-38 Lightning. Fire ate through the metal sheath of his right engine, melting its metal, thirsting for the gasoline that sloshed at Dave's feet.

On any other day, the turbo-supercharged engines would have been a thrill to throttle. The Lightning had assembled a lengthy résumé of firsts. She was the first fighter designed with twin engines, the first to carry rockets and bombs, and the first to fly over Berlin after the United States entered the Second World War. In February 1945, when President Franklin D. Roosevelt requested aerial protection for his flight to the Yalta Conference in Crimea to meet with Churchill and Stalin, he was escorted by none other than P-38 Lightnings.

In the skies above the Pacific, the P-38 was setting records. She was the first Allied plane to fly reconnaissance missions over Tokyo and also the first to land on Iwo Jima. The P-38 shot down more Japanese planes in the South Seas than any other fighter. Like a ninja in the night, the Lightning could swoop down over battleships, unleash hell, and then vanish into thin air. Even on one engine, the Lightning could square up against Japan's Mitsubishi A6M Zero.

After V-J Day, the Army Air Forces shed light on the mysterious death of the notorious Isoroku Yamamoto, commander of the Combined Fleet of the Imperial Japanese Navy and mastermind of the Pearl Harbor attack. Instead of dictating the terms of peace from the White House, as Yamamoto had arrogantly announced he would do, he was instead struck and killed by a squadron of P-38 Lightnings or, as the kamikaze called the warbird, *ni hikōki ippairotto*: "two planes, one pilot."

The Nazis had a name for her, too—*der Gabelschwanz Teufel*, "the fork-tailed devil." Luftwaffe pilots could readily identify its bifurcated rear stabilizers. During aerial combat, they rarely confronted the devil head-on. Unlike the P-51 Mustang that shot from the wings and struck targets effectively only at specific distances, the P-38 crammed all of its armament into its central fuselage. When the plane blew its nose, it produced a quintuple stream of fire that included four Browning .50-caliber machine guns each containing a maximum of five hundred rounds. The machine guns were staggered to create more space for larger ammunition boxes, and in the center of the cluster was the ever-lethal 20mm cannon. When its trigger was pressed, pilots could reliably strike targets from ten feet to ten football fields away. In North Africa, seventeen Lockheed warplanes downed twenty-seven Luftwaffe without losing a single pilot. In the Aleutians, a single P-38 sank a submarine. German tank commanders kept an eye out for the fork-tailed devils, which were known for harassing and targeting the sixty-ton vehicles.

But now, in the skies over Greece, Dave became the target.

Four hundred feet of flames trailed behind his cockpit, a comet of smoke and fire. Another damaging round of flak struck Dave's cockpit, shattering his canopy and peppering his instruments. Beams of sunlight poured in through the holes where his gauges once were. One engine still functioned—the P-38 pilots called it the "round-trip home." But Dave's fuel lines were leaking and the fire was spreading. It was only a matter of time before the Lightning lived up to her name and struck the Adriatic Sea.

The night before, Dave and his flight buddy, Joe Weber, had grabbed the D-flight jeep and hit off for town right after chow. The night was pleasantly cool. Fields of wheat and olive trees flanked the seven-mile drive to San Severo. The heart of the town was an officers' club, the only one in the district. It was a great place for coffee and doughnuts, and if you looked hard enough you could spot a pretty woman or two.

Joe squeezed his jeep through the medieval-sized streets, passing by baroque buildings bombed out by the Allies. They drove to the west side of town to check on their laundry, which for fifty cents a bundle would be washed by a bevy of beautiful Italian women. Hopefully, Queenie would be there. No GI had ever successfully dated the young Italian laundress, despite her proficiency in English and German. But that didn't stop them from trying.

After a good-natured rebuff from Queenie, Dave and Joe headed for the canteen. The night was still young, and the club wouldn't fill up for hours, so they drove to the cinema to catch a film. It was an old picture, one that Dave had seen several years earlier back in the States, but at least it gave the boys something to do.

After the movie, they raced like mad back to the club, which by now had become unusually crowded, and grabbed a few cups of coffee. The table next to them was occupied by a noisy bomber crew who looked entirely out of place in their winter flying jackets. Dave and Joe batted the breeze with them, comparing notes as all Air Corps pilots did. As it turned out, they were Americans with the Eighth Air Force and had just flown in from Russia, and they were making a big deal about it.

Joe spoiled their fun with news that the Fifteenth Air Force had already flown a mission in Russia. The bombers denied it, but everyone at the table was able to agree on two fundamental truths about Russia—the women were rugged, and the vodka was potent. About that, there was no misunderstanding.

Dave and Joe left the bomber boys to their bragging and started back for Triolo, giving a ride to an Indian soldier along the way. With no muffler on their jeep, the trio made quite the racket, bounding through the bombed-out town.

Like most fighter pilots, Joe enjoyed the speed. Only six weeks earlier, he'd survived a plane crash over the coast of Yugoslavia. Now, on the dark Italian roads, he stepped on the gas, feeling as invincible as ever. The little mass of nuts and bolts rolled at a good clip, reaching seventy miles an hour with a cloud of dust in its wake. Each time they hit a hole on the shell-pocked pavement, the jeep went airborne.

The driver was speeding too fast to see the object in the road. But Dave saw it. He shouted, and the jeep swerved, screeching to a halt and almost plowing into a shell hole. The three men jumped out to find an American GI lying on his back in the road. He was conscious but clearly in shock. His legs had been shattered below the knees. What remained of his arm hung loosely in his shoulder socket, barely connected to its torso. The man was trying to say something when Dave spotted the overturned jeep on the opposite side of the road. Its wheels were facing upward, its engine, smoking. A weak voice came from beneath the carnage, crying out for help.

The three men rushed to the wreck, hoisted the jeep with everything they had, and then removed another injured soldier. His head was split wide open and a jagged piece of skull protruded from a gash in his blood-matted hair. Dave heard another noise. He flicked his cigarette lighter and followed the sound through the darkness to a nearby ditch where he found a third victim, an American stretched out in the dirt with a pool of bloody saliva frothing at his mouth. The man's chest was turned inward, his rib cage crushed like a matchbox that someone had stepped on.

The pilots couldn't just let him die. They had to do something. There was an airfield about four miles away, so Dave climbed in the jeep with his new Indian pal and raced off, leaving Joe behind to offer what little first aid he could.

The speedometer reached fifty, sixty, and then seventy miles per hour. Aside from his headlights, the road was pitch black. His turn was fast approaching, and Dave threw the steering wheel wildly to the left. His back wheels fishtailed into a ditch but crawled out again as Dave forced the engine forward. The jeep tore down the dirt road, accelerating faster. Dave didn't see the railway tracks hidden up ahead, and he struck them with a bone-jarring impact that catapulted the men into the air and nearly flipped the jeep.

Somehow, they arrived at the San Severo airfield. The poor Indian guy bailed out, refusing to ride any longer. Dave located an ambulance crew, and the medics jumped into their rig and followed the jeep back to the accident, struggling to keep up.

By the time they arrived, a small crowd had gathered around the wounded Americans. A fight broke out between Joe and an eager colonel who insisted on moving the GI with the crushed chest. The colonel threw a punch at Joe and Dave tackled him, knocking the man into a ditch. An MP captain took it from there, leaping toward the rogue colonel and making short work of him.

"It's a good thing this fellow wasn't moved," one of the medics told Dave, looking up at him from the man's outstretched body. "His back is broken."

With nothing more they could do, the two pilots returned to Triolo Airfield. They arrived around midnight, still jittery, their muscles beginning to ache from lifting the overturned jeep. Dave threw away his blood-soaked clothes, wrote a letter to his family, and then hit his sack. He was grateful to be on the ground tomorrow.

October 6, 1944

At 5:30 A.M. the next day, the deafening roar of B-17 engines woke Dave from his slumber. *Those poor boys*, he thought with a grin. He rolled back over to snatch a few more hours of shut-eye. At 9:30 A.M., he tumbled out of the sack, left his tent, and inhaled the crisp Italian air. Wearing only his flying boots and a T-shirt, he strolled to the latrine. The sun lit up his fire-red, disheveled hair. It was a beautiful day in Triolo, and Dave was in no hurry. Nor were the clouds that skimmed lazily across the elysian sky.

An officer intercepted Dave. "Mac, we've got to have you," he said. "We can't find Nerney, and we need you to fly."

As it turned out, the Fourteenth Fighter Group's commanding officer had put in a rush call for four pilots from Dave's outfit. With Al Nerney missing, Dave rushed back to his tent and threw on his khaki uniform. It was cold in October, so he slipped a one-piece worsted wool gabardine uniform over his khakis before climbing into his flight suit. He was still tying his GI shoes when the jeep pulled up to drive him to the briefing.

Dave took his seat among the other twelve P-38 pilots. In

the center of the briefing room was a large map with colorful strings stretched across it representing their flight paths.

"Is that crazy Brezas on this one?" someone asked, looking around.

He wasn't. Top ace and flight leader of the Forty-eighth Fighter Squadron, Michael Brezas was somewhere else that day. Leading the squadron was First Lieutenant Donald Luttrell, a West Point graduate who had been a flight instructor in the United States and only recently assigned to the wing. He had the rank, but no experience.

Fifteen minutes later, Captain William Edwards walked in. He apologized for the mix-up and explained that headquarters had a last-minute change of plans. Captain Melvin Cooksey delivered the pilots' assignment—a strafing mission in northeast Greece. The primary target was the Salonika-Sedes airfield, a little more than three hundred miles away.

Since the German invasion of Greece in April 1941, the grass-covered airstrip near the village of Sedes had become the primary base of operations for the Luftwaffe. It was the perfect transit stop for German aircraft attacking Athens. Eight miles northwest of the Sedes airfield, on the fringe of the bay, was Salonika (also known as Thessaloniki), the second largest city in Greece. At the beginning of the twentieth century, half the population of Salonika was Jewish, but by 1943, in hopes of making the city *judenrein*—"clean of Jews"—the Nazis had killed 96 percent of its Jewish residents, deporting some forty-eight thousand people out of its train station. Most died in the gas chambers of Auschwitz a few hours after they arrived.

Now, in the autumn of 1944, the Allies were pushing the Germans out of Greece. The Nazis were on a full-scale retreat, and Salonika was their final stronghold. It was an ancient city, founded in 315 BC and named Thessaloniki after princess

Thessalonike, the half-sister of Alexander the Great. Their father, Philip II, had devised a brilliant military tactic called the *phalanx*. It allowed infantries to move across the battlefield in tight formation, simultaneously striking the enemy with pikes and defending their comrades with shields. Before he was thirty years old, Alexander used his father's innovative maneuver to conquer most of the known world. Dave could not have known how crucial the ancient war tactic was about to become.

The mission was a squadron effort and called for thirteen P-38s to fly in formation toward the Salonika-Sedes airfield. They were to approach from the bayside, coming in hot and low, but avoiding the tripwires placed near the water by the Germans to prevent such strafing attacks. The Lightnings would seek out targets of opportunity and inflict as much damage as possible on the evacuation planes parked on the airdrome. Dave scribbled down a few notes as the floor opened for questions. How many flak guns and where?

Second Lieutenant Royal Gilkey, the Officer of the Intelligence Section, didn't know for sure. Probably six or eight, he guessed. Reconnaissance planes had failed to capture a clear photo of the target area, so the exact flak positions were unknown. The weather specialist confirmed that the maps, too, were unreliable. The pilots would only be given general headings.

Was there a backup plan? An escape procedure? No. If a pilot got lost, or separated, or shot down, he'd be out of luck on this run.

Dave leaned over and whispered to Joe, "Bet you ten bucks you'll feather up and abort before we hit the Alps."

A few pilots laughed.

"Olly, come home with your wings intact this time, huh?" someone blurted out.

Four weeks earlier, Oliver Bryant had dinged up his P-38F during takeoff, and his buddies wouldn't let him forget it. Of course, it could have been worse. At least he hadn't completely destroyed his ship like Dave's tentmate Paul Forster did on September 4. Just exactly how "Frosty" walked away from that crash was anyone's guess.

"You'll probably be wise to carry your .45s today," the CO said, sobering the banter. "Everyone check your Mae West and dinghy. We'll be on the deck over the Adriatic coming back." If the pilots went down over water, they wouldn't stand much of a chance without their inflatable life preservers and rubber rafts.

The CO checked his watch and cut the meeting short. "Well, that's it, boys. Takeoff time is 10:53 A.M. for the Forty-ninth, 10:59 A.M. for the Forty-eighth, and 11:05 A.M. for the Thirty-seventh. You can take off for the flight line now."

The flight surgeon chimed in. "Good luck, boys. I wish I were going with you."

Like hell you do, Dave thought.

The whole operation was a rush job. The pilots made a mad scramble for the flight jeeps, raced to the "abort" to relieve themselves, and then back to the shack to gear up. Dave grabbed his leather gloves, earphone plugs, oxygen mask, and helmet.

It was a bright day, so he put a Polaroid filter in his goggles. He donned his dog tags and his identification bracelet. The pins in his chute needed to slide free if he was forced to bail out, so he examined them closely. He also gave his escape kit a proper check, making sure it contained food rations, a compass, and a handful of other necessities—all of which, Dave knew, would have to be hidden or destroyed if he went down over enemy territory.

He stashed his Ray-Bans in his knee pocket, along with some gum and Life Savers. Barney handed him his maps, and

Dave snagged his knife, pistol, and Mae West before surrendering his wallet, fat with $210.00, to Gilkey.

Dave handed his flight boots to his tentmate. "Take these, will ya, Frosty? I'll pick 'em up when I get back." He gave his chute to Johnny, his crew chief, and headed for the flight line.

The truck drove sluggishly down the airstrip as the pilots hopped off one by one at their ships. About halfway down, Dave disembarked and rendezvoused with his crew chief, who was arranging the chute beside the P-38.

The plane was only three months off the assembly line. It belonged to John Barnes Thomas, a pilot in Dave's Forty-ninth Fighter Squadron, and was so new that its serial number, 44-24131, had yet to be obscured by the exhaust of the supercharger.

A large, black *35* was painted on the plane's nose. Above it were three circled swastikas, badges of honor denoting John Barney's victories. The pilot had added the name of his sweetheart, Joanne, to his right engine cowling. After the war, Joanne Otto would return the favor. She would marry him and add his name to hers.

Not every P-38 of the Second World War was as romantically adorned, particularly the *Wicked Woman*, the *Vivacious Virgin*, and the *Beautiful Bitch*. Those planes belonged to a different class of pilot. Philip M. Goldstein, a highly decorated pilot in the Forty-ninth Fighter Squadron, painted *Jew Boy* on his plane, a name that was bound to arrest the attention of the Luftwaffe he downed.

As it turned out, *Joanne* was the latest model in the Lightning series, the P-38L-5-LO, which was the upgraded version of the P-38J that Dave had spent twenty-nine hours flying the month before. The U.S. Air Force produced 2,520 of this particular model, and there wasn't a need to ship the plane across the Atlantic Ocean. The Lightning could fly to Europe all by herself. With the installation of two drop tanks, filled with three hundred U.S. gallons each, the P-38L could travel an astounding three thousand miles nonstop. She became the ideal long-range escort.

The upgrades on the P-38L virtually erased previous mechanical problems. Five-inch High Velocity Aircraft Rockets (HVAR) could now be fitted beneath the plane's wings. In one modification, three-tube bazooka launchers were mounted to the reinforced pylon stores, or shackles, beneath the wings. Even more impressive, the P-38L could now carry four thousand pounds of bombs—more than the B-17 Flying Fortress could carry at the beginning of the war.

Even though she was one of the heaviest fighters in the air, pilots described the Lightning as one of the "daintiest" planes they'd ever flown. With the hydraulic aileron booster system, a single pilot gained the strength of ten men working the controls. For increased stability in the empennage, adjustable balance tabs were added to the rudders. The braver flyboys took full advantage of the upgrades by performing daring maneuvers only twenty-five feet off the ground.

In level flight, the L model could hike up to 425 miles an hour. In a power dive, upwards of 575 miles an hour was possible. At this great speed, pilots no longer had to fear the loss of control due to the paralyzing effects of air compressibility because Lockheed had mounted dive flaps under the wings, which acted like air brakes and allowed pilots to recover out of the fiercest of plunges.

The speed of the P-38L was unthinkable. Each of the twin Allison V-1710 liquid-cooled 12-cylinder engines stabled 1,425 horses of power for takeoff, and 1,622 for maximum boost. For pilots holding the reins during combat, the upgraded power transformed eighteen thousand pounds of plane into a light-footed ballerina, spinning and twirling through the skies. *Joanne* was sure to keep Dave on his toes.

Unlike single-engine fighters, which were cheaper to produce, easier to maintain, and more comfortable to fly, the P-38 kept her pilot busy. There was always something to do, something to adjust or fidget with. The Lightning never let you relax. Later on, she would be overshadowed by the P-47 Thunderbolt and the P-51 Mustang, but for many aces flying in World War II, the P-38L was the ultimate plane in a league of her own. "You just couldn't get away from the P-38L," said Arthur Heiden, a Lightning pilot in the Eighth Air Force. "Whatever the German could do, the American in the P-38L could do better."

Agile at every altitude, the Lightning maneuvered at forty thousand feet the way most planes maneuvered at sea level. Rising at a rate of 4,750 feet per minute, the "homesick angel," as pilots called her, wasted no time climbing into the clouds. Major General Paul B. Wurtsmith, commander of the Thirteenth Air Force, described the Lightning as "an instrument of destruction that works as well 1,000 miles from base as it does 10 miles away."

Dave examined the exterior of his new ship. He walked around the propellers and their red nose tips. He checked the tricycle-shaped landing gear and its wheels. The Fowler flaps looked fine. So did the central pod, or nacelle, which resembled a Venetian gondola.

Dave climbed up the boarding ladder and stood on the wing. From his vantage point, seven feet off the ground, the Lightning looked like a small mountain of steel and aluminum. Painted on the fins and rudder were blue horizontal stripes that distinguished the Forty-ninth Fighter Squadron from the pilots of the Thirty-seventh, which had red stripes, and the Forty-eighth, which had white. On the surface, the P-38L was impressive, indeed. Deep down, though, Dave knew she was hiding her quirks. Every plane had secrets.

A knot tightened in Dave's stomach. Something about the day felt off. Dave wasn't even supposed to be in the air. His mind still loitered around the horrific events of the night before. Had the three GIs survived? There was also something green about the replacement officer who would be spearheading the squadron into combat. He was a West Point graduate, and well-credentialed on paper, but Dave knew he lacked practical combat experience.

Equally unnerving was the fact that Dave didn't have time to retrieve the good luck charm he always hung from his gunsight reticle. The little statuette somehow boosted his morale, strange as it seemed to him. It gave him a sense of security, a calm in the storm, swaying there in front of his eyes. To every Lightning pilot belonged some lucky relic—the stocking of a wife, the rattle or boot of an infant. In the heat of combat, a man needs reminding what he's fighting for, what he'd lose should he fail to return.

By itself, any one of these changes was enough to rattle most fighter pilots, who, like sailors, can harbor a superstitious side. But taken together, they were ominous omens of a mission hexed.

Johnny, the crew chief, pulled the chute up the wing and helped Dave into the harness. Dave crawled into the cockpit and slipped on his gloves. Johnny tightened his shoulder straps. It was getting hot and sweat began to bead inside Dave's helmet. He cracked the primer, checked the coolant shutters and turbo levers, then chewed the cud with Johnny, who was sitting on the wing.

A few minutes later, the crew chief hopped off. Dave pulled the top panel down and locked it, hooked up his radio, flipped the main switch on, and held the energizer while hitting the starter switch. The blades turned over, throbbed, then graduated into a smooth roar. Now the other engine. Soon, the airfield erupted with the sound of thirteen Lightnings coming to life.

The engines had a certain rhythm in them, a pulse you could feel. Like a mother attuned to her infant's cry, every Lightning pilot knew when to feed the mixture, when to raise the rpm, when to give the aircraft a little extra attention. The cues were swaddled in the sound, pulsing from the plane's twin racing heartbeats.

Dave listened to his engines and gave them a minor run-up. The hundred or so major instruments all seemed to be functioning. Several Lightnings taxied out of the protective revetments. Dave waited for three of them to pass. He was assigned as the fourth pilot—*Tail Ass Charley*, as they called it. With his left hand on the throttle quadrant, Dave kicked off the brakes, gunned his right engine, and turned the plane to the left.

Taxiing the P-38 was exhausting. The short pedals wore out a pilot's shins, and before long, Dave was feeling it. A dust cloud poured into his cockpit as the winged caravan came to a standstill. He slipped on his Ray-Bans, hitched up his oxygen mask, and checked the temperature of the cylinder head. It was scalding hot.

"Step on it," Dave called out over the radio. Unless he got airborne soon, Tail Ass Charley would be forced to abort and cool off.

Somewhere behind him was Anthony Jerasonek, a Lightning pilot in the Thirty-seventh Fighter Squadron. Shortly after takeoff, he would encounter difficulties, crash on the runway, but live to tell about it. Anthony was one of the lucky ones.

The Lightnings lurched forward in a slow processional, and Dave finally got his turn. He rolled up his side windows, adjusted the tow throttles, then revved his engine until the turbos cut in. He glanced over his right shoulder to his wingman. Then he flipped the salvo switch on his left just in case he lost an engine during takeoff. The last thing Dave wanted was to crash with a belly full of fuel.

Joanne bucked at the brakes, eager for the gun, listening for the gun. There was nothing to do but wait. Two years of training had brought Dave to this moment. It was 12:05 P.M. on October 6, 1944, and the start of his fifth flying mission in World War II. The twenty-year-old was determined to make each second count, come hell or high water. Or both.

CHAPTER 10
STRAFING
SALONIKA

"Number 35, cleared for takeoff."

Dave advanced the throttles to forty-five inches and released the brakes. The Lightning shot out of the gate. In eleven seconds, it reached 105 miles per hour. Dave pulled back on the control wheel as the nose leapt off the runway. He raised his gear, raised his flaps, eased back on the throttles to synchronize with his wingman, then safetied the salvo switch. Together, the four Lightnings in Dave's outfit entered into a gradual turn some two hundred feet above the airfield, goosing their ships into formation behind the squadron leader, Donald Luttrell.

Dave's flight leader, Lloyd A. Bro, called for the check.

"One, okay."

"Two, okay."

"Three, okay."

"Four, okay," Dave reported over the crackling airwaves. "All okay back here, Bro."

The squadron leader at the front pulled out his maps to set a course. Dave's wingman was having trouble. Both of his fuel tanks suddenly fell off—he had forgotten to safety his salvo switch.

"Abort," Dave called.

His wingman complied, broke formation, peeled away, and returned to Triolo Airfield.

By the time the three P-38s of Dave's outfit joined the rest of the squadron, they were sailing at twelve thousand feet and well on their way above the Adriatic Sea. Bro pulled out his maps, double-checking the reference points.

Damn good flight leader, Dave thought. But there was a problem. Bro's wingman couldn't get his gunsight working. For some reason, it had gone dark. Without a spare bulb, he was forced to abort.

The situation had become a bit humorous to the remaining pilots. A mask covered Bro's face, but his oxygen hose shook from side to side. Dave could see that the flight leader was laughing in his cockpit. The four ships in the outfit had dwindled down to two. It was now up to Dave and Bro, *Joanne* and *Gloria*, to do all the shooting.

When the squadron reached the coast of Yugoslavia, they were met with overcast skies with a cloud base at ten thousand feet. The P-38s began crisscrossing back and forth, weaving in and out, providing cover for each other. Like the Greeks of old, they became a flying phalanx, moving as one, ready to attack, ready to defend.

Suddenly, Dave saw a long line of smoke coming from Bro's plane. His turbos were misbehaving. Bro fell out of formation and Dave followed to give him cover. Soon, the two planes were lagging ten miles behind the rest of the group. Dave double-tapped the radio transmitter button with his thumb to call up Bro, but before he could say a word the dogs started barking.

The unmistakable *woof woofs* of antiaircraft artillery filled

the air. Four black puffs blossomed outside Dave's cockpit.
Bro saw the 44mm shells go off and began violent evasive maneuvers, crisscrossing, pitching, dodging the incoming flak.

Twelve more rounds ruptured. Dave banked to see the source
and relayed the enemy positions, but Luttrell held course, resolved to fly in a straight line all the way to the target.

At the briefing, the pilots had been instructed to attack the
airfield from the bayside, flying low and avoiding the tripwires. But the mission wasn't going to plan. They were taking
a different path, over the mountains and toward the city of Salonika.

"Hold on to your belly tanks," Luttrell crackled over the airwaves. "Don't salvo them until I give the order."

The planes hugged the hills, dropping altitude as the landscape tapered off. The grounded Germans saw them coming and
littered the sky with orange bursts of flak. The squadron fanned
out in echelon. There was confusion. Some of the planes dipped
into gullies, others rose to higher altitudes. The attack was disorganized. The Lightnings lost their element of surprise.

A few seconds later, they were over Salonika. Luttrell forgot
to give the command to salvo, so a pilot in the Thirty-seventh
Fighter Squadron, John Jones, took charge.

"Drop your tanks," Jonesy ordered.

Dave hit his release button, but his tanks didn't budge. He
pressed it again. Maybe he was flying too fast? He checked his
airspeed indicator—four hundred miles per hour. Dave knew
he had to shake off the six hundred gallons of fuel. They were
as dangerous as carrying live bombs. Dave reached down between his legs and pulled the manual bomb-drop tank release.
Nothing happened.

As a last resort, Dave squeezed out a few rounds from his
cannon. Maybe the vibration would do the job. It didn't. The
tanks held fast, glued to his belly, waiting to detonate.

Dave's stray shots landed beside a Greek farmer who dove into a ditch for cover. The man's horse reared up beside him and fell over, twisting its attached plow. Dave barreled by, only twenty feet off the ground. He was close enough to see the fear on the farmer's mustached face.

A tree line approached. Bro fired a few shots into a cluster of mounds, thinking they were ammo dumps. Dave saw a dark object in the woods. He fired his machine guns. Something exploded. Flames coiled up around his windscreen as the wings of a German Ju 52, Hitler's darling plane, burned in his rear mirror.

Dave was now soaring over the Sedes airfield, the primary target. Both of his fuel tanks fell off, dislodged by the explosion of the German plane. The canisters dropped, skidded across the runway, and bounced into the headquarters complex. A mushroom cloud rose. Dave checked his airspeed—380 miles per hour. Everything was happening so fast.

In two quick beats of his heart, Dave had spanned the length of the airfield. He banked to see ten Nazi soldiers. The gunners had abandoned their batteries and were bolting across the field. Dave pursued them, descending like an eagle on its prey. He opened his guns and showered the men with quintuple streams of .50-caliber rounds. Their bodies bounced on the ground and looked like rubber balls in front of a water hose.

The sky erupted with the protest of fifty German guns spewing shrapnel into the air. The Sedes airfield was usually outfitted with only three batteries, but to Dave it seemed as if the Nazis had amassed all the artillery in Greece and were launching it at his fork-tailed devil. Hell broke loose.

Two pilots in the Thirty-seventh Fighter Squadron went down near the shoreline. At 2:25 P.M., Jonesy descended into a steep dive, opened his canopy, and hit the silk. A nearby pilot saw it happen. "I dipped my wing and saw his chute open," he later said, "and by that time I was too far past to see the plane or Lt. Jones hit the water."

Jonesy's wingman, William Milton, had also taken flak. He was last seen feathering his prop toward the north end of the bay near Salonika.

Dave was in trouble, too. He felt his left engine explode as scalding shards of metal, like red-hot nails, zipped up and between his booms.

"Mac, pull up. You're on fire."

Dave looked over his shoulder at flames flowing from the wing. It looked like blood gushing from an open wound.

"Shut up, Bro," Dave replied, spotting that his flight leader was also in flames. "Bail out!"

Flak peppered the two planes. Dave swiveled his head. The bursts were coming from the top of a nearby building where three German soldiers were shooting, their muzzles winking between each shot.

Dave still had one good engine, so he peeled away, dropped his altitude, and aimed his gunsight just beyond the soldiers. His right index finger found the trigger. Dave waited. The cannon begged. Suddenly, from the nose of the plane came a furious trail of 20mm rounds. It clobbered the three-man crew and minced the flak battery. The building collapsed.

Dave zoomed over the target and pulled back. His body felt six times heavier under the crushing force of gravity. Then he muscled the control wheel forward, went weightless, and pitched the plane straight down into a full-throttle, wide-open, guns-blazing nosedive. Dead astern were two German planes. They exploded. Dave leveled out thanks to his dive flaps. Flak came up at him and shattered his canopy. His gunsight completely disappeared, blown right off. Another target came into view—a Dornier flying boat. Dave aimed at its fuselage and punctured the plane with bullets.

Dave still had plenty of ammunition, but his Lightning was coming undone. Tiny dots of fire spread across the nacelle and matured into flames. The plane's control surfaces weren't responding. Dave looked to his left to see the engine cowling flaking away piece by piece. A large gaping hole pierced his wing, and through it he could see the blue water of the bay. The fuel lines were exposed and leaking. He considered feathering his inoperative engine, which would cut off its fuel supply and keep *Joanne* from bleeding out, but there was no time for a tourniquet. Dave had one card left to play.

To bail out of a P-38 Lightning, the pilot had two options. He could turn the bird upside down and let gravity pull him out. Or he could roll down the side window, push off his seat, slide onto the wing, drop below the empennage, and then deploy the chute. Pilots were hesitant to crawl onto the wing and fall between the booms because their chute cords could snag the torpedo-looking elevator mass balance, or bob weight, that stuck out beneath the horizontal rear stabilizer. If this happened, the pilot would go down with his ship.

Both options were risky, but one thing was certain. When bailing out of a P-38, the pilot needed at least five hundred feet of altitude for the parachute to fully deploy. Bail out any lower, and you couldn't guarantee the "silk elevator" would open.

As fire spilled into his cockpit, Dave remembered the long list of names posted to the board at his outfit back at Triolo, names of pilots who'd burned to death because they had decided to stick it out too long in the cockpit climbing for altitude. Dave was determined to keep his name off that board, but it was getting too hot to climb any higher. Fuel was spewing at his feet. Fire was coming through the air duct. His cockpit was becoming his coffin.

*   *   *

Within the sixty seconds of 2:39 P.M., three P-38 Lightnings struck the Bay of Salonika. Lloyd Bro was flying at an altitude of four hundred feet. He bailed out, deployed his chute, and somehow landed in the waves unscathed. Richard Arthur Schmidt was not as fortunate. The twenty-two-year-old lieutenant of the Forty-eighth Fighter Squadron attempted to bail out at an altitude of fifteen hundred feet. He managed to open his canopy and flip his P-38 upside down, but no chute was seen to deploy. When his plane crashed into the bay near Salonika, Schmidt joined the half-million Americans who had already lost their lives in the Second World War. He was buried in an English cemetery in Mikra, Greece, and the U.S. Air Force posthumously awarded him the Air Medal for his heroism.

It was now Dave's turn to bail out. He checked his altitude but was only two hundred feet off the water, less than half the altitude his chute required. The odds of survival, he knew, were about three hundred to one. Dave unfastened his safety belt and smashed his fists into the canopy release. It swung back and jettisoned into the wind. He hoisted his 120-pound chute over the side window and pushed himself off the seat. But something was restraining him, holding him down, choking him. His shoulder straps were wrapped around his neck and he was flying at four hundred miles per hour. The noose tightened. If his chute accidentally deployed, the force would snap his neck. The wind would tear his body apart.

Dave reached through the hole in his seat, yanked at the frayed straps, and finally broke free. He pulled back on the control wheel, hoping for a stall. But at 350 miles per hour, the plane angled up, flying too fast to come down.

Dave had to get to the wing. The side window was in his way, so he tried to lower it, but the hand crank was jammed. He drove the window down with all his strength, hammering it

The MacArthur family, c. 1926.
Vaughn (age 26), Dorothy (age 23),
Dave (age 2), and Gene (infant).
*From the personal collection of
Charles R. MacArthur.
Courtesy Carol Ansel.*

he MacArthur family. *Left to right:*
ene (age 17), Vaughn (age 44),
orothy (age 41), Dave (age 19),
nd Charlie (age 15) on the day
f Dave's graduation from flight
aining at Eagle Pass, Texas,
arch 12, 1944. Lieutenant
olonel Vaughn H. MacArthur
elivered the graduation address.
*om the personal collection of
avid W. MacArthur.
ourtesy Sharon Kinne MacArthur.*

Nineteen-year-old Dave
with his Fairchild PT-19
at the Spartan School
of Aeronautics, Muskogee,
Oklahoma, fall 1943.
*From the personal collection
of David W. MacArthur.
Courtesy Sharon Kinne
MacArthur.*

Nineteen-year-old Dave
as an Army Air Corps cadet,
February 1944, shortly before
his graduation from flight
training in Class 44-C
at Eagle Pass, Texas.
*From the personal collection of
David W. MacArthur.
Courtesy Sharon Kinne
MacArthur.*

Second Lieutenant David W. MacArthur receiving his wings from his father, Lieutenant Colonel Vaughn H. MacArthur, following Dave's graduation from Eagle Pass, Texas, March 12, 1944.
From the personal collection of David W. MacArthur.
Courtesy Sharon Kinne MacArthur.

Dave climbing out of the cockpit of his P-38 Lightning at Triolo Airfield, Foggia, Italy, September 1944.
From the personal collection of David W. MacArthur.
Courtesy Sharon Kinne MacArthur.

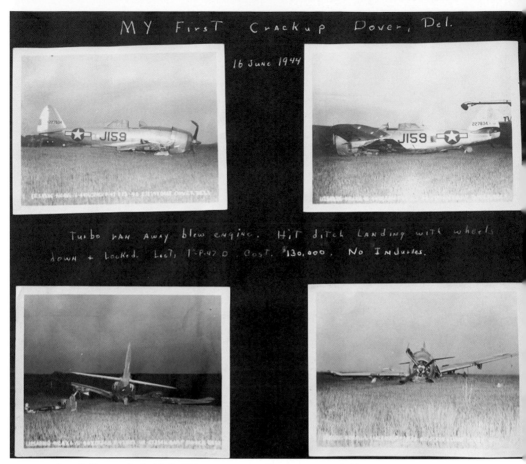

MY First Crackup Dover, Del.

16 June 1944

227834 — J159

227834 — J159

Turbo ran away blew engine. Hit ditch Landing with wheels down + Locked. Lost, 1-P-47 D Cost, $130,000. No Injuries.

Dave's first crash, in his P-47 Juggernaut, at Dover Army Airfield, Delaware, June 16, 1944.
From the personal collection of David W. MacArthur. Courtesy Sharon Kinne MacArthur.

Joanne, the P-38 Lightning Dave flew on the day he was shot down over Salonika Bay, Greece. John Barney Thomas *(left)*—who named the plane after his fiancée, Joanne Otto —and his crew chief. Likely August 1944. *From the personal collection of John Barney Thomas. Courtesy of his children, John B. Thomas III, Jeffrey Thomas, and Carolyn Thomas.*

The photo of Second Lieutenant David W. MacArthur taken for his escape kit, 1944.
From the personal collection of David W. MacArthur. Courtesy Sharon Kinne MacArthur.

'he pre-printed *Postkarte*)ave sent to his mother on Jovember 9, 1944, from the)ulag Luft interrogation camp ear Wetzlar, Germany.
'rom the personal collection f David W. MacArthur. 'hoto courtesy R. C. George.

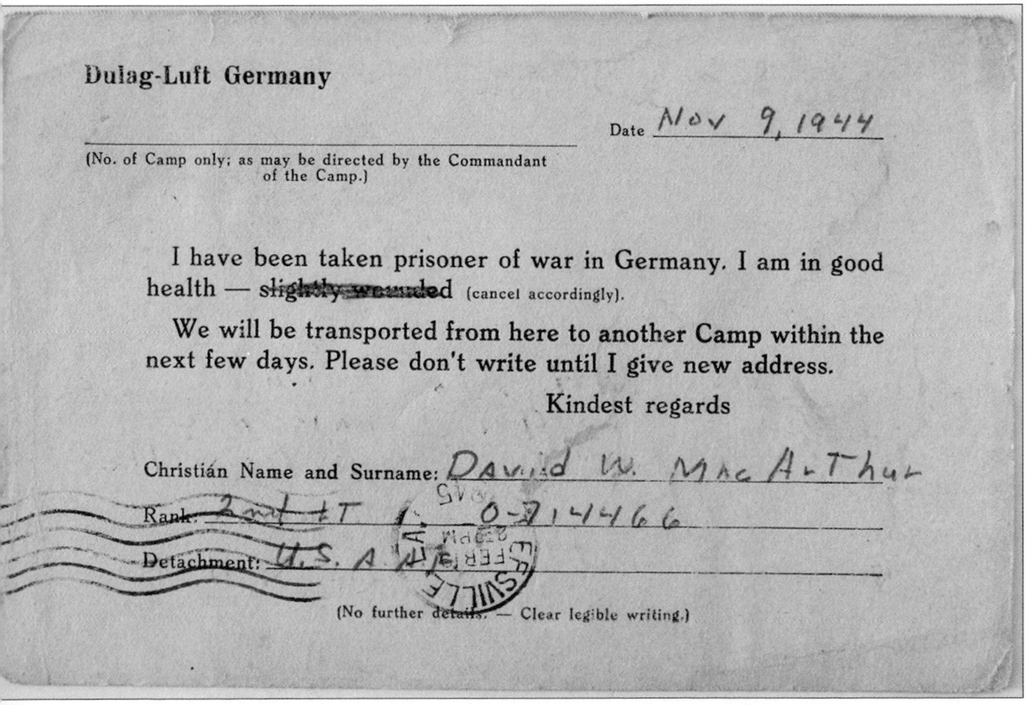

Dulag-Luft Germany

Date Nov 9, 1944

(No. of Camp only; as may be directed by the Commandant of the Camp.)

I have been taken prisoner of war in Germany. I am in good health — ~~slightly wounded~~ (cancel accordingly).

We will be transported from here to another Camp within the next few days. Please don't write until I give new address.

Kindest regards

Christian Name and Surname: David W. MacArthur

Rank: 2nd LT 0-714466

Detachment: U.S.A.

(No further details. — Clear legible writing.)

Left: Vaughn and Dave enjoy spaghetti and meatballs at Major General John M. Devine's house in Göttingen, Germany, May 4, 1945. *Right:* Dave and Vaughn on the terrace of Major General John M. Devine's house in Göttingen, Germany, May 4, 1945.
From the personal collection of David W. MacArthur. Courtesy Sharon Kinne MacArthur.

A staged promotional photograph, printed in the *Armored News* on June 4, 1945, showing Vaughn handing Dave soap and clothes in Göttingen, Germany, May 1944.

First Lieutenant David W. MacArthur *(standing, second from left)* leads a jeep convoy of men to freedom during the Fifth Chinese Offensive, April 22, 1951, Republic of Korea. Most of the men pictured would be killed moments after this photograph was taken. Several bullet holes are visible in Dave's jeep windshield. *From the personal collection of David W. MacArthur. Courtesy Sharon Kinne MacArthur.*

First Lieutenant David W. MacArthur, reading a letter in his radio jeep a few days prior to the bugout, April 1951, Republic of Korea. *From the personal collection of David W. MacArthur. Courtesy Sharon Kinne MacArthur.*

Above, left: Brigadier General Daniel C. Doubleday, the commanding general of the Rome (New York) Air Development Center, pins the Distinguished Service Cross on the chest of Lieutenant David W. MacArthur, Rome, New York, February 1952
Above, Right: Dave's Distinguished Service Cross.
Left: Lieutenant Colonel David W. MacArthur, United States Air Force.
From the personal collection of David W. MacArthur.
Courtesy Sharon Kinne MacArthur.

with his fist. Then he pushed himself out of the cockpit and
shinnied onto the left wing, his stomach directly above the reserve fuel tank. Dave made his body as flat as possible, but the
wind gave him a fierce smack in the face. Flames rose up and
around his flight suit, darkening his goggles. He had to hold
on.

Joanne was now losing momentum and approached her stall
speed of 105 miles per hour. Like a feline relaxing after a
spine-arching stretch, the Lightning crested and then came
down. Dave had been briefed about the bob weight on the horizontal stabilizer, so he turned his body all the way around.
With the wind at his back, he inched to the back of the wing
and slipped face-first down between the booms. He tumbled
head over heels, missing the bob weight, as the P-38 spiraled
out of control.

Dave felt frantically for the D-ring. At last he pulled the
wire and the chute strung out, but it refused to uncoil. Dave
plummeted to the water. *What words do I know in German?* he
wondered. Only one came to mind: *Gesundheit.*

Just then, the chute popped open. His body swung once and
slammed into the sea at a forty-five-degree angle. The icy waters flooded Dave's flight suit, filled his arms and legs, and
stole his breath. The impact ripped the goggles from his face.
His helmet was gone, too.

Dave sank quickly and struggled to escape from the parachute cords tangled around his arms. The more he fought, the
tighter the cords became. Everything was blue. Saltwater set
fire to his eyes.

Dave yanked two knobs on his Mae West and was instantly
propelled upward to the surface. He gasped. Just then, there
was a terrific explosion three hundred yards away. His P-38
Lightning, now a mass of fire, had struck the bay.

Seconds later, the water around him erupted with machine-gun fire. He'd been spotted by a gunner on the dock about a hundred yards away. Dave's parachute jerked and jumped as the bullets riddled its fabric with holes.

Dave freed himself of the cords, slipped off his chute harness, undraped his shoulder straps, and paddled for his life as effervescent lines cut into the water. He had to get away from the chute. Suddenly, the bullets stopped as the machine gun took a breath. Dave swiveled his head to get his bearings. He'd landed close to the shore, no more than seventy feet from the last plane he'd shot up. The twelve-engine Dornier flying boat, the evacuation plane, had been clearly damaged by his bullets, but Dave felt disappointed that it wasn't burning.

I'm a pretty poor shot without my gunsight, Dave thought.

The sinking aircraft diverted the Germans long enough for Dave to swim back to his chute and pull the dinghy out of its attached container. He twisted the CO_2 bottle and the raft inflated. He flipped it right side up and attempted to crawl inside. Buoyed by his Mae West, he heaved himself up onto one end of the raft, but the other end popped up and smashed him in the face with the CO_2 cartridge. Dave tried this three more times, frustrated that he'd never worn a Mae West in his twenty or thirty training runs. Finally learning his lesson, he reached over his head and pulled himself backwards over the side of the yellow dinghy.

In the distance, Dave saw a German boat heading out to retrieve Bro, who had landed about a quarter of a mile away in the water. It wouldn't be long before a vessel came for him, too.

Dave pulled his escape kit from his flight suit pocket and laid out its contents. The impact of the crash had bent and twisted the equipment, but he knew he had to destroy it completely. Dave ripped off his squadron insignia, saturated his identification papers, and tried to rip the maps apart. The maps were made of cloth and wouldn't tear, so he yanked out his .45, tore it down, and wrapped the maps around the disassembled pieces to dunk it. The weight of the weapon sank the maps to the bottom of the bay.

The German vessel was coming for him. Dave shoved the compass into his mouth, slipped a rubber-coated paper file into his shoe, and slashed the dinghy with his knife. The raft let out a long sigh but didn't sink. Dave released the blade, watched it shimmer downward beneath his legs, then looked up. He was staring at the wrong end of a German rifle pointed straight at his head.

CHAPTER 11

"FOR YOU, THE WAR IS OVER"

On October 6, thirteen pilots set out for the strafing mission over Greece. By 4:39 P.M., only eight of them had returned. During the attack, the Germans suffered the losses of three Ju 88s, a Ju 52, a Dornier flying boat, a Heinkel He 111, and two unidentified planes hidden beneath camouflaged nets. The Lightnings also damaged the three-story headquarters building on the Sedes airfield, two workshops on the north end of the runway, and countless flak batteries.

The next day, Lieutenant Luttrell, the West Point graduate and flight squadron leader, typed out the missing air crew reports for Dave and Bro. "We carried out the mission as planned," he stated. "I saw both men bail out and both chutes open. They were over the Bay and must have landed in the water. I could not stay to see where they landed because of flak in the area."

Dave was listed as a POW. He was last spotted at precisely 40° 32′N 22° 55′E. No attempt was made for his search or rescue.

* * *

Dave choked down a few gulps of seawater as the boat's waves washed over his head. He kicked frantically in the water, struggling to stay upright in his skinny Mae West. His limbs were growing fatigued, and there was blood in the water.

Four angry-looking Germans towered over him, likely the crew from the Dornier flying ship. They aimed their weapons at his head, shouting incessantly.

Dave squinted up at their dirty, yellow, burlap knickers. *They look pretty stupid*, he thought. The corners of his mouth began to turn up, but he restrained the smile. If the Germans saw the compass stuffed in his cheeks, they'd be liable to shoot him.

Two Germans leaned over, grabbed Dave's arms, and pulled him roughly to the deck. A bayoneted rifle was shoved into his side. Dave felt the blade penetrate his flight suit and sink a good inch into his flesh. Adrenaline numbed the pain, but warm liquid spread over his torso, ran down his ribs, and soaked his pants. The next blow came to his stomach, courtesy of the butt end of a German machine pistol. Dave nearly swallowed the compass and sucked vainly for breath.

Another rescue vessel pulled alongside and the guards tossed Bro into Dave's boat. The two Americans were thrown to one side and the soldiers watched them suspiciously from the other. Dave deflated the remaining air in his Mae West and exchanged a few mumbled words with his flight leader. Did Milton and Jonesy go down? Luttrell? How many ships did we hit?

A young German kid waved his bayoneted weapon at them, threatening to strike if the conversation continued. The Nazis' anger came as no surprise to Dave. They were very unhappy that he'd damaged their escape mechanism and killed a bunch of their buddies.

The rescue craft circled in the water to retrieve Dave's chute, or what was left of it. Three panels were missing, and the whole thing was full of holes, but somehow it was still afloat. The soldiers bundled it up and then followed the trail of burning fuel to the spot where *Joanne* had exploded and sunk. Other than the oily bubbles snaking to the surface from the submerged Allison engines, all that remained of the P-38 Lightning was a large green oxygen bottle bobbing gently on the waves.

Dave made a cocky comment, and the young German kid came over and jabbed his rifle into his ribs. Other crew members joined the assault. A fist caught Dave in the face and others kicked him mercilessly all the way back to the wharf.

Between blows, Dave managed to steal a glimpse or two at the airfield burning in the distance. The sight alone was analgesic. Glorious flames engulfed the dock and the aircraft near the runway. The squadron had completed its mission. At about 3:15 P.M., the boat arrived at the dock. The prisoners were greeted by two guards armed with machine pistols, a large crowd of women and soldiers, and four or five Nazi staff officers who were jabbering away in guttural German. Dave's dog tags had ripped from his neck in the crash, but he noticed that his identification bracelet was still fastened around his wrist. Maybe his name would show up in the list of Allied POWs.

One of the officers came over to Dave and asked him, in broken English, if the other pilot in his plane had survived. Evidently, the officer didn't realize that the P-38, with its two engines and twin booms, wasn't a two-man ship. Dave worked hard to keep the officer's ignorance intact. When the question was raised again, this time with greater force, Dave hung his head in sorrow, pretending to be sad over the loss of his imaginary copilot. The officer directed the same question at Bro, who assumed an equally crestfallen disposition. The ruse worked. The Germans smiled with pride at the might of their antiaircraft guns.

At that moment, Dave made a decision. No matter what lay ahead, he would make the best of the situation. If he couldn't strafe the Nazis from the sky, at least he could deceive them on the ground. Dave resolved to spend all his energy and creativity in thwarting the plans of the Germans. He would use his captors instead of letting his captors use him. Dave's attitude of stubborn resilience would endure for the entirety of his captivity.

A frenzied crowd tore apart the tattered remains of Dave's chute as the two Americans were taken off the wharf and placed in a makeshift jail where they were split up and thrown into different holding cells. There was a window and a bunk in Dave's room, but no latrine. It had been seventeen hours since he'd last urinated, and the strafing mission had kept him too busy to use the relief tube in the cockpit.

"You stupid baboon," Dave called to the guard. "Where can I take a leak?"

A large, stocky German barged into the room with an MP 40 Schmeisser submachine gun strapped over his shoulder.

"For you," he shouted, shoving Dave against the wall, "the war is over!"

The German tore off Dave's wet flight suit and undergarments, leaving only his shoes untouched. Then, to the great pleasure of the crowd gathering outside, he slammed Dave flush against the iron-barred window, exposing his naked body for everyone outside to see.

Dave felt like a zoo animal on display. The cold muzzle of the submachine gun pressed into his back as dozens of German women came up to the window, gawking with amusement at the nude American. Most of the women were prostitutes who had been imported to Greece, a sorry lot of ladies, Dave reckoned. The women were dressed in shabby, cast-off military uniforms once worn by the Nazi staff officers they serviced. Some of the women whistled sarcastically, laughing and hollering. Dave could feel their eyes roaming over his body. A few of the women marshaled their saliva, came right up to the window, and spat.

At least they've got bad aim, Dave thought.

The embarrassing experience lasted an entire hour, and Dave was determined to make it count. He tore his eyes away from the onlookers and surveyed the damage on the airfields. The flyboys had done a damn fine job.

The Germans scurried like bees around their crippled aircraft. Dave could see a twin-engine pontoon ship in sad shape near the docks. Another plane was being towed to shore by a rescue craft. Dave remembered squeezing his 20mm cannon at it and watched with delight as the injured bird struggled to stay afloat. Dave savored every second of its demise. The wing collapsed, the nose dipped into the water, and then the fuselage began to sink. It was better than watching a film, and he had a front-row seat. When the tail vanished beneath the waterline, Dave calculated that a total of ten Luftwaffe planes had either burned on the airdrome or sunk in the sea.

To his left was the field itself, now covered in fire and smoke. A series of ammo dumps was exploding, one after the next, and Dave counted each explosion. By the time he reached sixty, the mayhem was mesmerizing.

A rescue craft suddenly pulled up to the dock. Dave saw another captured P-38 pilot step onto the wharf. The poor chap looked pretty miserable. A blood-soaked bandage was wrapped around his head, covering his eyes. As the soldiers marched him up the wharf and near the window, the face became recognizable. It was Milton. The women lost interest and the amusement wore off, so the crowd dissipated.

There was little time to waste. Night was coming, and Dave

knew the guards would give him a thorough body inspection. He had to hide his compass in a more discreet location. He removed the compass from his mouth and shoved it up his rectum. A few agonizing seconds passed. Then Dave removed the rubber-coated file he'd hidden in his shoe and inserted it behind the compass.

At dusk, a German doctor entered the cell and began the physical examination. He slipped a glove onto his finger and into Dave's mouth. Then he instructed Dave to bend over, spread his legs, and pull his cheeks apart.

The Nazi didn't speak English, so Dave communicated with hand gestures. If the doctor wanted Dave's cheeks spread, he could damn well spread them himself. The doctor understood perfectly. He shrugged his shoulders and walked away, content to truncate the examination.

Night had fallen, and with it came a terrible chill. Dave was still naked and trembling when one of the guards opened the door and tossed him a mangy horse blanket. The fabric was filthy and smelled of manure, but desperate for warmth, Dave had no choice but to wrap himself in the rough wool.

Within two minutes, Dave's whole body was red-hot and covered in a rash. His skin felt the pricks of a thousand tiny needles. The deep wound in his back, lanced by the German bayonet, was itching for attention, but there was something else coming off the blanket. The cell was dark, illuminated only by the dim glow of the airfield burning through the window, but Dave could see the army of tiny red bugs that were eating him alive. Once content to inhabit the horse blanket, the menacing creatures were now moving in a mass exodus over his body—on his neck, in his hair, beneath his armpits, across his chest. Millions of them, it felt like, discovered a home between his legs.

Dave still hadn't taken a leak. The overwhelming pressure

on his bladder had to be relieved, so he yelled for the guard, motioning through the small window in the door, saying the only German word that came to mind. The guard just smiled at him. Very well.

Unable to hold it any longer, Dave walked to the corner of his small cell. An eternity later, as he turned to walk away from the wall, he realized that between the water from his clothes and his urine, the floor was covered. He sloshed to his bunk, the filthy blanket wrapped around his shoulders and the smell of urine filling the room. He eased gently onto the edge of the bunk as the outside fires extinguished, leaving his room in pitch darkness.

Sometime later that night, Dave would be dragged away by the same brutal German who had forced him to stand naked at the window. The *Hauptmann* would throw him a pair of baggy pants and a bloodstained tattered sweater. Dave would scarf down a few stolen meat rations before being crammed into a Volkswagen with Bro and Milton.

But until then, in the agonizing minutes and hours that remained, Dave lay on his back in the bunk, legs straight out and stiff. The sharp file in his bowels threatened to tear him internally. Even the slightest motion brought waves of pain.

Every once in a while, he dared to swat at the bugs trespassing across his body. But for the most part, Dave remained perfectly still, straitjacketed by his suffering, alone and tormented in the dark.

CHAPTER 12
NO USE PLAYING POSSUM

In the dead of night, a Volkswagen careened down the narrow Greek roads that coiled around the Bay of Salonika.

"Where are we going?" Dave asked. The driver didn't respond, but the question did elicit a response from the German captain who faced backwards in the front seat. In his hand was the machine pistol that Dave, Bro, and Milton had become overly accustomed to seeing.

The Hauptmann explained that he'd studied English for five years in school, then proceeded to test a few jokes on the pilots. No one laughed.

The three Americans felt weak and hungry sitting in their returned wet flight suits, so right in front of the garrulous guard Dave removed from his pocket some of the iron rations he'd stolen back from his emergency kit. The stash amounted to little more than a few tablets of malted milk, dextrose, and benzedrine, the first-generation pharmaceutical amphetamine. It wasn't much but it was better than nothing.

"You like this car?" the German asked, looking at Dave,

who had apparently become the designated spokesman for the group.

"It's nothing new to us," Dave bragged. "Your buddies left plenty behind in Africa and Italy."

The Hauptmann snorted with laughter.

"Have you ever ridden in one of our jeeps?" Dave asked.

No, he hadn't, but he'd seen pictures.

The Volkswagen approached a closed gate, and the small talk ended. The driver exchanged a few words with the officers outside, the gate rose, and the vehicle pulled into a courtyard that was partially encircled by a building as modern as any Dave had seen in the States.

The three Americans were taken through the entrance and up several flights of stairs. They arrived at a large suite of offices and a spacious laboratory the Germans had confiscated from the Greeks. Lab tables were all piled in one corner of the room, and there was a handful of cots stacked against the wall. In the center of the room was a wide desk with a light suspended above it.

As they entered, a half-dozen high-ranking staff officers broke into a dramatic tableau of salutes. It was an obnoxious eruption of "Heils" and handshakes that seemed to Dave like something straight out of a Nazi propaganda film. The German captain was tasked with gathering the Americans' names and serial numbers, but Dave had no intention of making it easy. He responded only with silence.

The captain, growing ever more impatient, grabbed Dave's identification bracelet still attached to his wrist and copied down the information. Then came the litany of questions, all of them asked in German. Dave laughed loudly.

The Hauptmann, ever eager to impress his commander, took Dave's disrespect personally and shouted at the Americans for a solid twenty minutes. Then he screamed something into a nearby telephone as Dave continued chuckling. A few minutes later, an old duffer with a lot of braid and not a hair on his head entered the room. He squinted at the Americans through the monocle slotted into his eye socket. Dave studied his uniform and surmised that he was an *Oberst*, comparable to a colonel, and now the highest-ranking man in the room. The Oberst spoke in German to the other officers, and something about the way he looked at the pilots concerned Dave.

"This fellow evidently speaks good English," Dave murmured to Bro, "and he doesn't want us to know it."

Dave turned back to the Oberst. "You can help us," he said, "and it's no use playing possum. We're very much on our guard."

"I guess you are right," the man agreed.

Dave snapped to attention with a flair for ceremony. "Lieutenant David W. MacArthur, 0-714466." He bowed stiffly. The display felt silly to Dave, but it did the trick with the Germans. The attitude in the room shifted, the staff officers now showing a bit of deference to Dave. Sensing this change, Dave made his first request of the monocled Oberst. "This man has a terrific gash over his eye," he said, gesturing at Milton. "Would you fix him up?"

The man called over a young major and told him to take a look at the wound. After more salutes and "Heils," the two left and reappeared forty-five minutes later with Milton's cut washed and stitched. The Oberst sat next to Dave and began a meandering interrogation, likely hoping his goodwill gesture had thrown Dave off guard.

"Look," Dave said, "I know what you're after, and you should know that I'm going to be careful. Now I will cooperate with you as far as I can, but I will not betray myself. I am in your hands, and I know the Geneva Convention fairly well. I don't ask much."

The German nodded and agreed to get right to the point. He had just come from the field the P-38s strafed, he said, and reported that while six Americans had been killed, the Germans had not lost a single plane.

"Then why can we still hear the explosions, even from here?" Dave scoffed. "I have eyes."

The Oberst finally admitted that the Americans might have slightly damaged twenty-one planes. Dave figured it was at least double that. Not bad for an outfit of thirteen ships. But the man refused to concede that any Germans had been killed in the mission. "We have the bodies of some of your men in a truck outside," the man bluffed. "I will show you." When Dave agreed to go, the man balked and changed the subject. He offered the Americans ersatz coffee, but the boys' empty stomachs couldn't handle the crude substitute. "What is the matter?" the Oberst asked.

"We haven't eaten all day," Dave said, "and we're used to real coffee." With those two words, *real coffee*, Dave saw something that resembled nostalgia flash across the old man's face.

"I wish I had some," the Oberst mused.

From his pocket, Dave pulled out a concentrated cube of coffee, the last of the rations. He handed it to a German orderly, who hurried away to brew it. When the man returned, the Oberst offered the cup to Dave. "You take it," Dave said. With a tinge of embarrassment, the man sipped the warm elixir.

"What do you do at home?" the Oberst asked.

"I collect souvenirs."

"What kind?"

Dave thought for a moment. "Guns," he lied. "What's in your holster? I'd like to add it to my collection."

The Oberst laughed and pulled out a palm-sized pistol. "This would do you no good here."

"I know," Dave agreed, "but if I ever got the chance, it

would be handy to have. No matter how many guards you put on me, they can't stay on alert all of the time."

"That is true," the Oberst replied, this time with a shade of seriousness in his voice. "But it would be useless to try to escape. You might be sorry if you try anything."

Dave's face turned to steel and he locked eyes with the German. "Not half as sorry as you would be."

The moon hung high over the Greek countryside as the covered truck transported the Americans to their next stop. There were few signs of life on the road, other than at the occasional control posts.

When the driver reached his destination, he came to a stop and the four armed soldiers disembarked with their prisoners. They walked down a trail that led to a small shack nestled far back in the woods. It had been forty-eight hours since the Americans had tasted anything that resembled real food, and the German soldiers knew it.

With his wingmen on his flanks, Dave sat down at the table between Bro and Milton. A jolt of pain shot up his abdomen and he was painfully reminded of the compass and file still sheathed below. The Germans dished out a generous selection of black bread, margarine, and liverwurst, which turned Dave's thoughts to the food alone. He devoured it in no time flat.

"When will the war be over?" one of the Germans asked.

Dave knew what they were up to. In exchange for food, the Germans expected the Americans to provide information about the war. Between bites, Dave fed them a long line of bull. He told them about how he'd crossed the Atlantic Ocean with a massive convoy escorted by several heavily outfitted cruisers and six battleships. Two German submarines were targeted and sunk.

"When we got to the port for disembarkation," Dave continued, "we had to wait for four days while they unloaded six hundred ships that were ahead of us."

The Germans listened, their eyes growing wider. The Americans continued spinning their fabricated tales.

"I need to go out and take a leak," Dave said.

A soldier accompanied Dave outside, monitoring him closely. For the first time since his capture, Dave toyed with the idea of attempting an escape. During his training, he'd been taught to resist the enemy at all costs. If captured, he must attempt to escape, not only because it was the best chance for his own survival, but also because the distraction may tie up enemy troops and open opportunities for resurgence.

Dave turned his back to the guard and considered his options. He could probably kill the guard, take his weapon, and then make off with the truck. But how far could he get? His German was pitiful; his Greek, even worse. Also, what about Bro and Milton? He couldn't leave his buddies behind. The idea began to lose traction in his mind, but one day, he assured himself, there would be an escape.

After the meal, and when the Germans felt satisfied with the information they had gleaned, they loaded the Americans back onto the truck and pushed off. The night grew long and nearly reached into the next day. The houses that dotted the stretch of road began to thicken in frequency. When they reached Salonika, Dave could feel the pavement turn to cobblestone beneath the wheels. In the center square of the town, on Tsimiski Street, was the jail. But after much deliberation between the Germans and the Gestapo Field Police, it was decided that the overcrowded prison could not accommodate the Americans. The truck was forced to continue on its trek.

It didn't take long for the driver to get lost. Not knowing what to do, he ground to a stop, jumped out of the vehicle, and wandered up and down a back street, pounding on doors and demanding directions from the residents he awakened. Dave saw the man exceed the threshold of his frustration, silencing a barking dog with a swift kick of his boot. He returned to the vehicle only to find himself lost again a few miles up the road.

By 3:00 A.M., the tortuous journey finally came to an end. At the base of Mount Chortiatis, the truck approached a two-story white building surrounded by multiple layers of twelve-foot-tall barbed-wire fences. The back of the vehicle dropped open and the Americans, weary and sleep deprived, climbed out. Two guards marched them across the front of the long building to a wooden gate covered in barbed wire. The flimsy gate wobbled open. Just how exactly the pitiful barrier kept even a single man from escaping the compound was anyone's guess.

In the stillness of the night, the complex took on a ghostly appearance. The area was absent of life and strangely quiet. The Americans walked through the gate. An Italian man stood a few hundred feet in the distance, holding a lantern that marked the location of the barracks. Dave could not have imagined, as he stumbled through the darkness, what horrors awaited on the other side of the door.

CHAPTER 13
SMOKEY JOE

The first assault was the smell—a pungent, eye-burning cock-tail of horse manure, open latrine and, oddly, fresh garlic. Dave felt the full force of the putrid odor as the Italian holding the lantern swung open the door.

Piles of dung reached a foot tall in places and were jealously guarded by thick swarms of flies and mosquitoes. The three Americans didn't know which of their senses to trust the least—their sight or their smell. In the notorious Greek con-centration camp of Pavlos Melas, their home for the next two weeks, both senses would alarm them equally. Once a Greek Army barracks, Pavlos Melas was converted by the Germans in 1941 into the largest detention center in northern Greece. Located in a suburb of Thessaloniki, and nestled at the base of Mount Chortiatis, the death camp became the primary transit stop for POWs bound for Europe.

Life at the camp was animalistic. The Nazis converted the Greek army cavalry stable into a large barracks and made no effort to clean the facility before replacing the horses with humans. More than a thousand prisoners were crammed into the animal quarters. The Germans knew that the Serbs hated the Croats and were always at each other's throats, so they housed the two groups together.

Without any toilet facilities, all the prisoners in the camp were forced to dispose of their waste against a small brick wall that emptied into a lone slit trench. Unsurprisingly, the latrine often overflowed, contributing not only to the odiferous fumes, but also to the outbreaks of dysentery and malaria that ravaged the camp. The resourceful Serbs gathered fresh heads of garlic from the outside fields and strewed them across the barracks to keep the bugs away.

Dave, Bro, and Milton were ushered into a side room to meet the *Oberfeldwebel*. He was a short, squat German buck sergeant, the second in command at the camp, and had bony legs and a dopey expression that reminded Dave of the foo-fighting Smokey Stover from the Sunday comics back home.

"Smokey Joe," as Dave would nickname him, hollered at the Americans in German, trying to appear gruff. Dave couldn't understand what he was saying, so he shrugged his shoulders and looked around the room. The small, filthy area had one window, a potbellied stove, a few pans, two cots pushed into the corner, and a tiny table that suffered from a broken leg. On the table was a lantern, the only light source. On the wall hung Smokey Joe's clothes and weapons—a burp gun, a machine pistol, and a Walther P38.

Dave wondered how difficult the P38 would be to steal, and he kept his eye on the pistol. Smokey Joe ranted and raved for a few minutes then ordered the Americans to be taken away to join the others. The Italian led them out of the room and up a flight of stairs. Dozens of sleeping prisoners covered the stair-case, their bodies strewn in haphazard angles. The Americans stepped around the dangling limbs, weaving their way three floors to the top. Bro stumbled over one man, but the prisoner just grunted and returned to his dreams.

At the top of the stairs was a door, which the Italian opened. Dave walked into the room. It was six feet wide and twenty feet long and had a window that was partially covered with a screen. Dave counted six men in the double-decker bunks that lined the sides of the room. A beat-up table was in the center of the room, and two more men—one in a dirty British army shirt and shorts, and the other in what was left of a Royal Canadian Air Force uniform—were stretched across the table.

In the dim light of the lantern, the black walls began to move, undulating in swirling waves. Dave squinted and realized they were blanketed by flies and mosquitoes, the most he had ever seen in his life. They swarmed the lantern and dimmed its light with their tiny bodies. The Italian lit a match and moved it toward the light, and the insects fled. A small fellow with a scraggly mustache, a Greek by the looks of him, stirred in the lower-right bunk. Stretched across the bunks on the left were four British Limeys: a sergeant, a pilot officer, an aerial gunner, and a navigator. A Canadian named Toddy dragged himself from the table to welcome the trio. Dave detected in his voice an Illinois twang that suggested he grew up close to the U.S. border.

Toddy pointed to the eighth man sleeping soundly in a top bunk. He was also a newcomer to the camp. Milton walked over to the bunk and gasped. It was Jonesy, one of the boys from the Thirty-seventh Fighter Squadron. Apparently, he'd also been shot down on the strafing mission. Jonesy, now wide awake, jumped down and greeted his comrades exuberantly, while Toddy sent the Italian away to scrounge up a few more mattresses.

In the meantime, the noises in the room had attracted an audience, a tough-looking crowd of Serbs who spilled into the doorway. Some were young, but you wouldn't know it by looking at their faces, which had been creased and aged by two years of confinement at Pavlos Melas. They welcomed the Americans with outstretched hands filled with gifts of food— the camp's most precious commodity. One of the Serbs handed Dave a big bar of Nestlé milk chocolate. Dave would later learn it was the man's six-month supply.

To further enliven the fellowship, Toddy pulled out a large, round loaf of black bread. It weighed nearly eight pounds and was covered in green mold. With the blade of a broken saw, he started hacking away at the bread. Next, he opened a can of English corned beef, guarding it from the swarming insects. With his fingers, he pulled out chunks of meat and placed them on the bread, then handed the delicacies to each of the three men. Dave bit off a piece of the moldy meal but couldn't swallow it. He gave the rest to a Brit, and the man scarfed it down whole.

By this time, the Italian had returned with one straw mattress that was riddled with all sorts of vermin. The guys discussed what to do. One of the Serbs at the door gestured that two of the Americans could sleep in their beds.

"It'll please the Serbs," Toddy said. "And besides, it's a good deal."

Milton was a bit leery of the idea, so Dave and Bro left him to the straw tick and followed the Serb into the nearby sleeping quarters. When Dave's eyes adjusted, he saw moonlight spilling into the room through the side windows and calculated that the barracks was about sixty feet wide and three hundred feet long. It contained no fewer than eight hundred prisoners, some of whom were playing cards at long tables down the center.

Most of the Serbs and Croats were asleep, their emaciated bodies crammed together in makeshift beds that lined the periphery of the room. No man had more than one blanket. Their beds, which were elevated five feet in the air, amounted to little more than primitive frames with chicken wire slung over. The Serbs had tried sleeping on the floor, but when they did, the rats that roamed at night would bite and gnaw at their fingers and toes.

Dave followed the Serb who offered his sack to the Americans. It turned out that he was the CO of the outfit, and he shared his bunk with no one. Instead of chicken wire, his bed had wooden slats, a lumpy mattress filled with straw, and three blankets. Through a sophisticated combination of hand gestures and rudimentary Italian, the Serb proudly told Dave that his bed was nearly bug free. Dave would soon learn why. He lifted the blankets and saw thousands of garlic bulbs piled onto the frame. The acrid smell nearly stifled him.

Another Serb walked over holding a canteen cup. Dave didn't recognize the liquid, but parched as he was, he took a sip. It was terrible. His stomach crept upwards, but he forced himself to swallow. He soon realized that if he wanted to survive, he'd have to learn to like the vile drink.

The two Americans peeled off their wet flying coveralls, removed their shoes, and crawled into the garlic-flavored bunk. Dave felt a sharp pain in his bowels. He curled up and carefully removed the rubber-coated file from his rectum. He knew it would come in handy for the escape he would plan. So would the compass, which was worth its weight in diamonds, but he couldn't find it.

I guess I'll get that out in the morning, Dave thought as he drifted to sleep.

For the next three months, Dave examined every bowel movement with scholarly attention, picking through his feces with a stick, hunting for the artifact. But in the end, the compass was not to be found.

* * *

A few hours later, sometime before 5:00 A.M., Dave was awakened by the sound of eight hundred prisoners abandoning the barracks. Evidently, it was their daily routine to rise before dawn and go to work for the Germans. Some labored in textile mills and warehouses in Salonika below, while others worked at the Sedes airfield.

After the Serbs all left, Dave rolled over to find a spot in the bed that wasn't full of lumps. Within minutes, he was back asleep. Three hours later, the sun opened Dave's eyes. Bro was sitting on the edge of the bed, shaking from the cold. He'd been up all night, unable to sleep. In the daylight, it became obvious that the poorly ventilated room had once been used to stable horses. The walls were full of holes and torn at joints where partitions once stood. The ceiling, some thirty feet above their heads, was bare and skeletal.

Dave and Bro collected their clothes, left the empty barracks, and retraced their steps to the small room to find Milton, Jonesy, and the others. The Brits were making their best effort at preparing something for breakfast, but Dave wasn't interested. He collapsed in a bunk, desperate for more sleep.

A few hours later, the door burst open and in walked Smokey Joe, waving his pistol and straining his voice. He was accompanied by another German guard, who motioned for the prisoners to follow. Dave grabbed his flying coveralls, which by this time were badly wrinkled but beginning to dry, and put them on over his German prison garb. The eight men followed the guards to the camp's quadrangle at the center of the compound.

The Greek army had once used the quadrangle as a drainage field, but to counteract the soggy soil, the Germans paved the area with headstones looted from the Jewish cemetery in Salonika. Before 1942, the cemetery was among the largest and finest in Europe. It contained half a million graves, most of them dating to the fifteenth century. After the war, the Jews returned from exile to discover the cemetery utterly desecrated. The Nazis used the marble headstones to pave roads, construct houses, and even decorate swimming pools.

Standing on the headstones was a long line of Serbs, exhausted from the day's work. Dave sized them up and was impressed by their clothing. They were finely dressed in black zippered jackets and corduroy pants donated by the American Red Cross. Square gray caps crowned their heads. The Serbs had just returned from the crippled Sedes airfield, and more than a few were grinning at the Americans. They had witnessed the effects of their strafing mission over Salonika, and to them, Dave and the other pilots were heroes for inflicting so much damage on the Nazis.

Dave surveyed the area surrounding Pavlos Melas. In the hazy sunshine, he saw the city of Salonika down the hill. Its buildings looked white as bone. Behind him were the mountains he had flown over in his P-38 Lightning the day before. A large field, about a thousand yards across, separated the prisoners' barracks from a tall, barbed-wire fence. Beyond the fence was a second set of barracks where the Nazi guards were quartered. Compared to the converted stable, the German barracks looked brand new. At the base of the mountain, Dave saw two 88mm flak battery positions, each containing three to four antiaircraft guns. He knew those guns all too well. So did *Joanne*, who was sleeping peacefully in her silty grave at the bottom of Salonika Bay. To the right of the guns was a second compound surrounded by three layers of barbed-wire fence. Even at a distance, Dave could see the starved inhabitants— women, children, and a few elderly men. They looked like corpses in tattered clothes. Dave's eyes jumped from one skeleton to another, squinting in disbelief. There were no young men. A long trench stretched between the compounds, filling the air with a stench carried by the wind. Dave saw a third compound surrounded by barbed wire. He didn't yet know it, but the compound contained roughly five thousand people, mainly Greek women and foreign internees who lived in dirty huts. Trails of smoke rose up from the compound, making the whole area look to Dave like a littered, burning garbage dump.

Loud German shouts snapped Dave back to the foreground. The Serbs popped their heads to attention as Smokey Joe and his staff strutted by. He was accompanied by two Italian soldiers and another German Oberfeldwebel with a bulging cyst on the back of his head. The cyst was so large that it forced his small hat forward, tilting it perilously over his eyes. The guards paced up and down the line of Serbs, counting the prisoners. The group was as large as an entire infantry company. Dave watched Smokey Joe lose count and start over seven times. He eventually gave up, pretending to arrive at his desired number. When he reached the Allied pilots, Smokey Joe noticed that Dave didn't spring to attention. In fact, Dave gave him the cold shoulder, refusing eye contact. The German was infuriated by the lack of respect. He turned his back, eyes wide, and stormed off to the barracks. Thinking the inspection was over, Dave also turned to walk away.

All of a sudden, someone hollered and Smokey Joe burst back, shouting angrily and waving his gun in the air. Dave held his ground, stubbornly refusing to acknowledge the fuming German officer. Smokey Joe was livid. "It is time to view the 'reprisals,'" he snarled.

A few minutes later, several German guards arrived from the second compound with about forty Greek women and children. The Germans marched them in front of the prisoners, stripped off their clothes, forced them to their knees at the edge of the long trench, and then raised their guns. Some of the children were crying, clutching the hands of their mothers.

Just then, the piercing sound of machine-gun fire split the

air. Women and children began to fall forward, their bodies tumbling into the pit. A few more bursts and everyone collapsed. The air went silent except for the pained shrieks of one young child. A bullet had shattered the girl's leg, but she was still alive.

One of the guards walked over to the girl and kicked her small body into the trench. She landed on top of her dead mother, her muffled cries rising from the pit. A Nazi guard walked forward with a sack of white powder and began emptying it onto the girl. The caustic lye burned her legs, fell across her face, and filled her mouth. She screamed louder as the powder began to burn her skin. The screams went silent as the guard turned the sack over, shaking the last of the white dust into the pit. Soon, the girl's young flesh would be liquified into a brown sludge, along with the other bodies in the trench.

Under the special orders of Field Marshal Wilhelm Keitel, signed on September 16, 1941, one hundred civilians were to be executed to atone for the loss of every German life. For every injured Nazi, fifty civilians must be exterminated.

On September 2, 1944, less than one month before Dave arrived in Greece, twenty trucks of German soldiers entered the nearby mountainous village of Chortiatis. The Nazis gathered the residents in the village square, raped the women and children, and killed many of them with their machine guns. One group managed to escape to a nearby house, but the Germans pursued them, locked the doors, and set the house on fire. Thirty-four civilians were burned alive.

The Nazis led another group of civilians to a bakery. The doors were locked and a machine gun was mounted in the bakery window. Against the backdrop of a violin, jubilantly played by one of the German soldiers, seventy-nine people were killed. At the end of the Chortiatis Massacre, the Nazis had burned three hundred houses and slaughtered 146 civilians. Among the dead were 109 women and young girls.

At Pavlos Melas, only eleven miles away, the same soldiers found no shortage of humans to sacrifice. The locals called the executions the *blood tax*. The International Military Tribunal of the Nuremberg Trials later called them *terror murders*. But to Dave, who witnessed the horrific ritual every morning for two weeks, it marked a turning point in his hatred of the Nazis.

This wasn't a war. This was a holocaust.

CHAPTER 14
SABOTAGE

October 7, 1944, Pavlos Melas, Salonika, Greece

After the Nazi firing squad carried out their morning executions, some of the Serbs meandered over to the Allied pilots, offered them cigarettes, and tried to teach them a few Serbian words. Dave was still processing the murders he'd just witnessed and looked at them half-aware, his eyes glazed over.

A few minutes later, the Germans marched the Serbs back to their animal stable. Dave and Bro followed to retrieve their damp clothes. They returned outside, walked to the edge of the compound, and hung their flight suits on a wire. Fifteen feet away was a warning wire, Toddy said. If anyone stepped over it, he'd be shot.

Dave glanced over at the German guard perched with his gun on his tower as he walked with the group to the top of a grassy knoll, the highest point in the camp. Far away from the stench of the barracks and the trench filled with decomposing corpses, Dave stretched out in the sun. It was the first time since his capture that he'd had the chance to hear how his buddies arrived at the death camp. Milton had been downed over the water, struck by enemy flak. His P-38 hadn't caught fire, though, so he feathered his inoperative engine to keep the propellers from windmilling. The landing gear was jammed. When the wheel touched the water at three hundred miles per hour, the plane bounced a few times. Without his shoulder straps fastened, Milton lurched forward and smashed his forehead on his gunsight, nearly losing consciousness. When his ship finally came to a stop, he scrambled onto the wing and was picked up by a nearby boat full of Greeks. Milton didn't even get wet. The Greeks wrapped a bandage around his head and hid him belowdecks, but one of the sailors turned chicken and signaled a German patrol boat. That sailor, Milton later learned, was hanged for treason by the Greek partisans.

When Milton arrived on shore, he saw Dave's chute still on the dock. It was covered in blood and riddled with bullet holes. The Germans told him that Dave had drowned in the crash and that they'd let his dead body sink into the bay.

Jonesy was in the same flight squadron as Milton. They'd been given the same mission—strafing the Megalo Mikra Airdrome about three miles south of the Salonika-Sedes airfield. Jonesy took some flak and his fuel lines burst. Gasoline sprayed into his cockpit. With the frayed electrical lines threatening to ignite the plane, Jonesy bailed out at about four hundred feet and landed on the airfield. A nearby flak battery came by to retrieve him. They loaded Jonesy into a car and took him to the Sedes airfield. It was there that he saw the damage Dave did to the headquarters building, which was still burning from his belly tanks. On account of all the smoke, though, Jonesy didn't see *Joanne* go down and was curious about the damage done to the other airfields.

Dave filled in the missing information. He described the exploding ammo dumps, the burning airfields, and the floating plane that sank while being towed to shore. There was plenty of time to watch the whole scene unfold, Dave explained. He got a front-row seat standing at the window in his "birthday suit."

Bro joined the conversation, explaining how he had named his right engine after his mother, Anna, and his left one after his wife, Shirley Mae. Right at the start, Bro had been struck with flak, but he didn't know it until one of the two "women" stopped speaking. "I was very hurt," Bro mused, "to think that a motor named after my wife would give up the ghost." Bro had seen the *Joanne* slide by, trailing smoke and flames, but it wasn't until Dave's timely warning to shut up and bail out that Bro decided to bail. He pitched the *Gloria* up, rolled her inverted, and bailed out at four hundred feet. He landed in the water about a hundred feet offshore. The Germans caught him while Bro paddled in his dinghy toward Dave.

Then Jeff spoke up. He was a short, British guy with a scraggly mustache who'd been the front turret gunner of a wimpy, twin-engine, long-range bomber based in Africa. He was a radical, a renegade who went AWOL and was court-martialed. While under arrest, though, he snuck out for a raid with his flight crew. That night, Jeff's bomber was struck by a Jerry night fighter. There was a blinding flash of light, and the next thing Jeff knew he was in the water. A German boat picked him up. He spent the next seven months in a hospital in Athens suffering from a mess of broken bones—both legs, his spine, a collarbone, and about six ribs. Jeff admitted that, at first, the Brits were suspicious of Dave, Bro, Milton, and Jonesy. They didn't take to them right away because they assumed the Americans were Germans posing as pilots to get information. But something about Jeff seemed suspicious to Dave.

Nick and Toddy had been on a strafing mission over Crete. Their plane, the impressive long-range Bristol Beaufighter, ran into the blast from one of their own rockets and crashed. The impact catapulted both men out of the shattered plane and into the water about three miles off the coast of Crete. Floating nearby was a chute. Nick and Toddy swam over to it, inflated its one-man dinghy, and then floated helplessly at sea for three days. The crash sliced Nick's feet, which swelled up twice their size and became infected. On the fourth day, the situation became critical. Without water or food, and at the mercy of the tide and winds, they ended up drifting to the only town for miles, which was occupied by Germans. Nick and Toddy joined Jeff in the hospital in Athens, and by the time they arrived in Salonika, the gangrene in Nick's foot started to heal.

Frank told his story next. He'd been a commando sent to Crete on special duty. Dropped behind enemy lines, he fought in the resistance movement on Crete. For two years, he and two Greeks blew up German airfields, killed as many Nazis as they could, and attacked enemy garrisons. Misfortune struck during a night raid when a German shepherd attacked Frank. A Greek resistance fighter shot the dog, which alerted the German sentries. They made a run for it, hijacking a car that belonged to a Nazi general, and then drove into the hills. The next day, one of the Greeks in their company betrayed them.

The Germans came in hot. They surrounded Frank, shot at him, and tossed grenades at his position. A blast went off, killing a Greek and filling both of Frank's legs with scrap iron. He was captured, taken as a prisoner, and handed over to the Gestapo, who tortured him sadistically. Among the many torture techniques practiced by the Gestapo, perhaps the most feared was the "water cure."

Frank was tied down and a water hose was inserted into his mouth. For two days, liquid was blasted into his mouth. It wasn't long before he lost his mind. A week later, and in a semiconscious state, Frank felt something begin to tighten around his face. His torturers had placed his head in a vise and were twisting it tighter. Frank felt the jaws of the vise squeeze his skull. Bit by bit, the pressure increased. Eventually, the pain became unbearable. Frank's eyeballs popped out of their sockets. The vise was loosened and his eyeballs recessed back into his head, but for the rest of his life Frank's sight would be impaired. After another week of torture, Frank was thrown into the Athenian hospital, where he met the other boys.

Dave's least favorite of the pack was Stan, the "yellowest individual" he'd ever met. Stan was the navigator on a Beaufighter when mechanical problems forced the plane down over water. He and his pilot scrambled into a dinghy, but they were captured by the Germans after Stan gave away their position and signaled the enemy in the dark. A group of Greek resistance fighters liberated the two Brits and killed the German crew. Five weeks later, Stan's pilot convinced the Greeks to help them escape. The plan was to sail from Crete to the mainland and then travel to Yugoslavia. Stan wanted only to surrender but tagged along reluctantly.

The evening after their departure, their boat needed fuel and pulled into an inlet. A German vessel approached and gunfire was exchanged. Stan hit the deck while the Greeks jumped overboard to swim for the shore. His pilot swam to safety, less than a hundred yards away, but Stan refused to go. He was captured by the Germans and by that time was so nervous, so out of his mind, that they sent him to the hospital in Athens to be checked out.

Perhaps the most interesting man, a fighter who was somehow mixed in among the eight flyboys, was George, a modernday Sinbad, by Dave's estimation. The language barrier was a problem because George spoke Greek and Arabic far better than English, but eventually the fascinating and mysterious details of his life came into focus. George was born in a whorehouse in Cairo, Egypt. His mother was a prostitute of Turkish and Arab ancestry, and his father was a Limey on a tramp steamer. When George came of age, he set out on his own adventures, taking to the high seas as a sailor. During the early days of the Second World War, he claimed to have occupied every ship in the world that could float. "And some that did not," George said.

One such ship, a British vessel, went down in the Mediterranean, and George swam for shore. For a long time, he wandered through Greece as a lone-wolf guerrilla fighter. The Allies paid him more, George explained, so he fought against the Germans. He was eventually captured, handed over to the Gestapo, and for three months endured all manner of torture.

Not only did he experience the infamous water treatment of having a hose shoved down his throat, but George was also waterboarded nonstop for three consecutive days. On the final day, the Gestapo beat his stomach so fiercely that he vomited blood. Several of his organs ruptured.

The guards then tore out his thumbnails and poured salt water on the raw, bleeding cuticles. They plunged his feet into pots of boiling water and squeezed his head in a vise so tightly that his skull cracked. Before they sent George to the hospital in Athens, he was tied to a wall and beaten into unconsciousness. It would be three weeks before George awoke from the coma, and six months before his mind finally returned. By then, he'd lost nearly a hundred pounds, and he'd also been castrated by the surgeon in order to save what remained of his penis. Even after that, after enduring the worst torture of any man in the group, George wore a wide smile on his face as he stretched out in the grass. Dave's respect for the man was enormous.

Sixty Allied prisoners had been hospitalized in Athens. When the Germans were being driven out of the city in early October, they shuffled the prisoners to Salonika. Fifteen men, including some of Dave's new buddies, were loaded into a boxcar and given one loaf of black bread to eat. The ration had to last for three days, the Germans said.

The 310-mile train ride up the Greek coast was grueling for the prisoners. Partisans had blown up nearly all of the bridges. With no time to rebuild them, the Germans ran separate trains between each of the blowouts. As each train reached a river, the Nazis unloaded the prisoners and forced them to wade through the water to the train waiting on the other side. The three-day trip stretched into seventeen. The loaf of bread was long gone.

One night, as the guards lay sleeping, four prisoners succeeded in prying open one of the boxcar windows. One by one, the men crawled out the window, jumped off the train, and ran for their lives. The Germans awoke, hunted down the escapees, and shot the men. As George was telling the story, Toddy leaned over to Dave. "Jeff and Stan are the ones who tipped off the guards," he whispered.

When the train finally approached Salonika, the Greek partisans anticipated their arrival and launched a furious attack on the Germans. The train squealed to a halt amid the whine and whir of bullets. Stan and Jeff were found huddled in the corner of the boxcar, whimpering.

Dave listened to his new friends share their stories until the day darkened. The sun had failed to dry the dampness in their clothes, so the men left them hanging on the wire and trudged back to their barracks. As usual, flies covered the walls. Everyone started to think about food.

"The Germans are bringing in some soup tonight," one of the Limeys said. Sure enough, the door eventually swung open and in walked a guard carrying a rusty, round bucket. When not filled with food for the prisoners, Dave learned, the container functioned as a garbage can for the guards.

Dave leaned over to survey the slop and saw large chunks of meat floating beneath at least two inches of golden grease. A frugal dusting of barley, no more than a pinch or two, spotted the surface of the stew. When the tin bowls were filled and the stew was divvied up, one of the Brits extinguished the lantern on the table. Dave thought it was odd, but his hunger overcame his curiosity and he joined the others for the feast. Dave poked his finger into the bowl. He could feel the cold layer of slime and the large chunks of meat surrounded with rice. Before long, Dave realized why the prisoners ate in the dark. The rice was wiggling.

The stench of the barracks disguised the smell of the soup. Dave lifted the liquid to his nose and inhaled a putrid waft of spoiled flesh. He walked into the hallway, borrowed the light of the one dim bulb, and stared down at his bowl. It was filled with rotten horsemeat from casualties of an earlier Allied strafe, along with maggots. Dave instantly lost his appetite, as did the other Americans who couldn't force a single swallow. But three days later, their hunger returned with such ferocity that they were glad to darken the lantern and scarf down the black and blue meat. *I'm probably going to eat worse*, Dave thought. *I might as well get used to it*. After a while, the slop didn't taste so bad. Even the plump, squirmy maggots became tolerable to chew.

Every morning was the same. Dave would be pulled from his barracks to line up with the Serbs in the quadrangle. Greek women and children would be stripped of their clothes and shot by the callous Nazi firing squads. Sacks of lye, too many to count, would be thrown over their bodies to aid the digestion of the trench. Never in his twenty years of life had Dave seen evil come so sharply into focus. There was only one thing at Pavlos Melas that promised to keep him sane.

After dinner each night, Dave would go to the window in his

room and gaze out over the city of Salonika. Perched on the third floor of the building, he was high enough to see a generous panorama of the area. His unobstructed eyes traced the ancient streets as they stretched down the hill to the docks, and beyond to the harbor. In the daylight, the city looked ragged and covered with filth. Its open sewers, war-torn ghettos, and bombed-out waterfront lay naked and exposed, unable to hide from the white light of the afternoon.

But when dusk fell across the bay, and the orange sun licked the city sideways, everything was baptized in golden rays of light. The dirty streets glistened like new. The dilapidated houses and slums looked entirely untouched by the wounds and weapons of war. Even the buildings and their bullet-blemished facades were spackled over by the warm colors of the sky. Through that bug-infested window, the whole city of Salonika, from its hills to its horizon, looked to Dave like a picture taken straight out of the Bible. It was a city reborn, a world putting itself back together before the death that came with the dawn.

At 5:00 A.M. the next morning, Smokey Joe burst into the room to count the prisoners. He did this often and unexpectedly, sometimes as many as ten times a day. The task seemed to empower him, to add some kind of meaning to the monotony of his days. Dave and Bro headed outside to retrieve their clothes still hanging from the wire. The night air had finished drying them marvelously, but something on the fabric caught Dave's eye. He pulled the flight suit off the line and noticed dark stains on the back of his pants.

The shock and adrenaline of the crash must have masked the pain in his leg, but Dave did remember being hit under the seat of his pants. Dave twisted his torso around and saw large red clots congealed to the back of his upper thigh. He ran his hand over the wound. There was something jagged and razor-like sticking out of his skin. After a few minutes of painful excavation, Dave removed two steel splinters, remnants of the flak that had exploded against the armor plate under his seat. He'd been hit worse than he realized. Dave showed his leg to the fellows, who summoned an Italian corpsman who played the role of camp medic. For the next several hours the medic excavated bits of shrapnel from Dave's leg with a sharp knife and tweezers.

Later that afternoon, the Serbs and Croats returned from their work with smuggled pieces of wood taken from the forest. They also had stashes of food and alcohol. Frank, the British commando, had taught them sabotage techniques while they were out working in the stores and warehouses of Salonika. After the Serbs learned to steal food from the docks, Dave and his buddies were eating well, and there was lots of booze to go around. The arrival of the Americans energized the Serbs. Soon, they were blowing up warehouses, looting the goods, and then splitting the spoils with the guards on duty. The goal was always to get the Germans drunk.

Keeping an eye out for the guards, one of the Serbs lit a fire in the washroom with the gathered wood. For the first time in days, Dave and the other pilots were able to shave with a contraband razor and boiling hot water, a luxury well worth the risk. As they were cleaning up, a Serb caught a glimpse of Dave's bleeding leg, broke off a piece of soap, and handed it to him. Dave lathered the soap between his hands, rubbed the mixture into his wound, and then poured the hot water over his leg.

His biggest fear was infection. The prisoners of Pavlos Melas were no strangers to gangrene, and many of them died without being treated with antiseptics. It took a full week of sitting in the sun and near-continuous rinsing before Dave's wound showed any signs of healing.

* * *

One afternoon, Dave was sitting outside in the sunshine when a Greek woman from across the fences offered to sew up his torn flight suit. Dave slipped the suit over the fence. The woman smiled and scurried back to her barracks as fast as her legs could carry her. Early the next morning, a German guard barged into Dave's barracks with the flight suit wadded in his hand. "You, come!" the guard shouted.

Dave's stomach turned sour as he followed the guard outside. In the cool morning air, Dave saw the woman who had offered to help him. She was now restrained by two German soldiers. She trembled in the quadrangle as they held her arms, stepped back, and then lifted their machine guns.

With cackling laughter that Dave would remember for the rest of his life, the Nazis cut the poor woman's body to pieces with the spray of their bullets.

The resistance at Pavlos Melas took many forms. With the exception of fresh food, cigarettes became the camp's primary currency. During the day, the Serbs looted the smokes from German stores. When the guards came up to Dave's barracks to trade tea and other goods, the prisoners would sneak behind the guards and filch valuables from their pockets.

One night, Dave lured Smokey Joe up to the barracks with the promise of a party. In ten minutes, he was drunk and began staggering around the room. He whipped out his Walther P38 pistol and pointed it in the men's faces, laughing and ordering them to drink up. It took all of three drinks to finish Dave off, but before the liquid lightning went to his head, he managed to steal Smokey Joe's pistol, bullets and all. The next morning, Smokey Joe awoke to his missing P38. He got plenty upset but dared not raise a fuss about it for fear he'd be punished. For the rest of Dave's time in Greece, he was in possession of the weapon, keeping it in his pocket, waiting for the right time and place to use it.

A turning point in the sabotage came one night when a prisoner stuck a cigarette down the barrel of a rifle that was slung over the shoulder of a German guard. It was a small act of sabotage, but it went unnoticed. Emboldened by their success, the prisoners decided to try out more daring maneuvers. Before long, Dave had crammed sand, nails, and all sorts of things into the Nazis' weapons. A few days later, the sabotage paid off. One of the guards went outside to clean his gun. He pulled the trigger of his rifle and was killed when the bolt blew back as the round exploded in the chamber. The Germans never caught on. Word of this technique quickly spread among the Serbs. Every afternoon presented new opportunities, especially at the Sedes airfield where they worked. They would wait until the guards were occupied and then load debris into the tank guns and flak batteries scattered near the runway. The next time they fired at Allied planes overhead, hopefully the vandalized antiaircraft guns would explode, injuring the gunners and possibly setting off nearby ammo dumps. For the Serbs, these were no small victories. Even from prison, they could fight the Nazis.

The ultimate sabotage, of course, the very thing every pilot dreamed of at night, was escaping from the camp. For nearly two weeks, Dave, Bro, and the others hatched elaborate plans for escape. Using loot stolen by the Serbs, they pooled their resources and bribed an Italian guard to cut a hole in the barbed-wire fence behind the camp. Dave sent one of the prisoners through the fence to contact the Greek partisans.

If the escape went according to Dave's plan, the resistance fighters would flash their signals from the hills and move into position, ready to attack the Germans if necessary. Dave and his friends would creep outside, make a wild dash for the fence, and then flee to safety before the guards could shoot them.

The plan was set and the night of the escape arrived. Right

on cue, the Greek fighters signaled the flash in the hills and positioned themselves around the barbed-wire fence. But just as
the men started off for the fence, Stan and Jeff panicked.
Afraid they'd be shot, the two men made a racket, hollering for
the Germans. The guards were alerted, the escape was foiled,
and Dave's first real chance at freedom was thwarted.

On October 19, as Dave neared the end of his two-week stay at
Pavlos Melas, Major General N.F. Twining, the Commanding
General of the Fifteenth Air Force, wrote a letter to Dorothy
MacArthur in Leesville, Louisiana.

My dear Mrs. MacArthur:

It is with extreme regret that I must confirm the report
of the War Department that your son, Second Lieutenant
David W. MacArthur, 0-714466, has been missing in action since October 6, 1944, when he failed to return from
a strafing mission to Salonika, Greece.

After leaving the target, your son was seen to gain altitude in preparation for a jump. His parachute was seen to
open. As this happened over Salonika Bay it is likely that
he was picked up by enemy rescue craft. I hope that we
will have more news in the future. Rest assured that the
War Department will notify you immediately should further information be received.

David was well qualified for his position as a fighter
pilot. He was respected and admired for his extreme
courage and devotion to duty.

In the name of his many friends here who miss him
keenly, I extend sincere sympathy.

CHAPTER 15
"HEIL JOE STALIN!"

The next morning, and only days before the camp would be liberated, a guard stormed into Dave's barracks with news. It was time to move. The Germans were planning their final retreat out of Greece, and word had reached the camp that the quick-paced Russians were advancing. The Allies were less than two hundred miles away. The prisoners hurried to gather their meager belongings. Dave layered his flight suit on top of his other clothes, tucked Smokey Joe's loaded P38 into his pocket, and took one last look at the window. Then he abandoned the small room and kissed its fly-covered walls good-bye.

The prisoners were herded to the gate outside the compound, where vehicles waited to take them to the Sedes airfield. There, the Germans forced them to help conceal some of the field guns in a nearby grove of olive trees, and as soon as they finished, a P-38 Lightning reconnaissance plane flew over the airfield taking photos. The ship received some flak, but the undermanned guns had little or no ammunition. Several of the 88mm rounds even exploded because they'd been sabotaged earlier by the Serbs. Dave spent two days on the airfield. Then, on the third morning, heavy trucks laden with German equipment pulled up on the airfield to take the American and British POWs away. Dave climbed atop one of the trucks at the rear of the convoy. It contained large barrels of fuel and seven German guards holding submachine guns. The trucks inched out of Pavlos Melas, passed through the crowded streets of Salonika, and then began the long northern journey out of Greece.

Somewhere along the transit, the road became clogged with columns of refugees. They were the saddest-looking humans Dave had ever seen. Old men and women were pulling carts designed for oxen. There were small children walking among the refugees. Some of the children had hitched rides on the carts while others trudged beside their parents with sagging eyes and faces as dirty as the road.

The trucks moved along the congested road at a snail's pace. Frustrated by the holdup, and trying to intimidate the prisoners, one of the guards jumped out of the truck, walked over to a refugee, and snatched an egg out of the basket in her hands. The woman resisted. The guard raised his gun. Then he took the entire basket and passed the spoils around to the other guards who began laughing and entertaining themselves by tapping holes in the eggs and slurping out the raw insides. Suddenly, a dog came running out of the crowd to chase the truck. Still laughing, one of the guards decided to use the dog as target practice. He aimed at the animal, fired, and missed. The bullet hit a refugee instead. The noise of the gunfire spurred the other Germans into a frenzy. They put down their eggs, turned their muzzles on the refugees, and began firing into the crowd. Some of the refugees collapsed under the bullets. Others didn't react at all. Dave assumed they must have been accustomed to being targeted by the Nazis. There were no loud cries, no acts of protest. The dead went silent. The wounded just lay on the ground, quietly whimpering in agony. Dave felt his blood boil over as he watched the guards picking off the men, women, and children just for the fun of it. His right hand was now in his pocket. His palm tightened around the grip of his stolen P38 and his thumb inched toward the hammer.

To save their bullets, the Nazis came out of their mania and returned to slurping their raw eggs. Dave's stomach turned against the Germans. He wanted to kill every single one of them. He knew he could put a bullet through at least one of their heads, but he also knew that the guards would immediately return fire and probably end up shooting all his buddies.

Dave weighed his options. Even with a pistol, a successful getaway seemed impossible, especially with Stan and Jeff in the truck. The traitors would squeal at the slightest provocation. A few more seconds passed. Dave relaxed his fingers. He pulled his hand from his pocket and let his arm fall gently to the side.

"If I thought that I could have done it successfully," Dave later said, "I would have shot every one of the guards."

The night came, and the trucks arrived at the assembly area. Dave had been whispering with Bro for the entire trip, occasionally glancing over at Stan and Jeff, who looked nervous. The convoy came to a stop, and the prisoners stepped out into the dark. When it was Dave's turn to jump off the truck, he exchanged a subtle nod with Bro and the two were about to make a run for it when, before they could bolt, the two Limeys cried out with a fearful warning.

The guards instantly encircled the prisoners, drew their guns, and marched the men away from the truck and through the woods. Dave reached into his pocket for his gun, but it was gone. He turned around to see if it had fallen out, but the barrel of the German weapon dug into his back, pressing him onward.

When they reached a small cottage, the Nazis kicked out the owners and ordered the prisoners into the house. Stan was first through the door and darted for the one and only bed in the room. Dave shook his head in disgust, still wondering about his pistol. He stretched out beside Bro on the hard floor. The night was filled with interruptions. Every five minutes, the guard burst into the room to count the prisoners. Sleep became impossible. At some point, one of the guards outside shouted a loud "Heil Hitler!"

"Heil Joe Stalin!" Dave hollered in response.

A staff officer burst into the room. With the Russians closing in, Dave's retort solicited a violent reaction from the German. He blew up, spewed out a string of guttural expletives, and then smashed Dave in the ribs with his gun.

An hour later, the guard came back in. "Eins. Zwei. Drei. Vier. Fünf . . ." The pilots, fed up with the constant interruptions, mutinied. Dave and Bro jumped to their feet, muscled the guard outside, slammed the door, and locked it.

"If you wake us up again," Dave shouted, "we're going to either beat your brains out or escape."

The guard wasn't seen again for the rest of the night.

Early the next morning, the Germans woke the prisoners and ushered them outside. Dave knew there was going to be hell to pay. Instead of loading them back onto the convoy trucks, the Hauptmann led the pilots to a ditch near the side of the road. He lined them up side by side, drew his pistol, and then ordered the men to turn their backs. Someone was going to die. Maybe they all would. The sun had just started to creep up over the horizon. Dave turned to face into the light and braced for the bullet. This wasn't the worst place to die, he reasoned, watching the Mediterranean sunrise. Behind him, Dave heard

the German's footsteps. He was pacing back and forth, decid-
ing whom he would shoot first. Twelve more guards approached.
In his periphery, Dave saw the firing squad take their positions. The Hauptmann pulled back the hammer of his pistol. A
few men flinched at the click.

This must be the end, Dave thought.

Not knowing which second would be their last, some of the
prisoners held their breaths. After a full minute of anticipation,
Dave heard a noise coming from the Hauptmann. He was laughing. The other guards were snickering, too. The execution had
been a hoax, an attempt to scare the prisoners and make them
feel powerless. The pilots had been awakened at the crack of
dawn not to die but, instead, to be transferred to another outfit
scheduled for an earlier departure.

For the rest of the morning, Dave felt the euphoria of having
survived. Every minute that passed was borrowed time. The
prisoners were placed under the command of a new Hauptmann, a man named Schultze who looked to Dave like the
quintessential high school teacher.

"Schultzy" surveyed his new prisoners as if he were buy-
ing a horse. He lined the men up, counted them, and signed a
receipt. For the next two weeks, these humans belonged to
him. Dave was marched to the assembly area, where sixty or
seventy other prisoners were being evaluated by a German
sergeant trying to look rough with his burp gun draped across
one shoulder and a bandolier of ammo flung across the other.
He looked out of place, like a middle-aged businessman turned
soldier. His English was good, so Dave warned his buddies to
hold their tongues. The sergeant split the men into groups of
four and loaded them into boxy trailers that were hitched to the
back of diesel trucks. Each trailer had a storage section that

could be accessed through the padlocked door on the back. Dave wondered what lay inside the secret compartment as he was led into the cramped cabin.

Four grown men were forced to share the tight quarters of the room, which was only five feet wide, six feet long, and four and a half feet tall. If Dave fully extended his arms and legs, he could touch all the walls at the same time.

Two short bunks, each only five feet long, gobbled up most of the room and were dressed with threadbare blankets that made no promise of keeping out the cold. Equally unaccommodating was the clunky stove in the corner, which had fallen into disrepair and served no practical purpose other than to occupy what little space was left.

Dave was unable to stand fully upright inside the trailer and had to hunch over. For a second, he was back in the claustrophobic cockpit of the P-38 Lightning for his piggyback ride. Three men were already sprawled out across the cabin, their limbs overlapping. Dave didn't know them, but he suspected that the journey would remedy the unfamiliarity.

Separating the driver from the prisoners was a tiny window. Over the course of an hour, as the prisoners were loaded into the trailers, a German face occasionally peered through the window. The guards outside were shouting loudly, and Dave passed the time picking out German phrases he'd learned. Eventually, the businessman-turned-sergeant entered the cab and the convoy lurched into motion. He didn't tell the prisoners, but they were bound for Bulgaria.

At a complete standstill, the tiny room was awkward enough. But when curving around bumpy roads that had been bombed by the Allies, the experience became nothing shy of toturesome.

Since the two bunks were already occupied, Dave sat in the wooden kiddie chair, which crumpled beneath his weight. The prisoners laughed. The sergeant scowled through the opening. As the truck picked up speed, the cool morning air spilled into the room through holes that were stippled into the ceiling. Dave collected the broken pieces of the chair, thinking that they could become firewood, and then he crawled over to the stove where the third man was sitting, shivering.

The convoy squirreled through the outskirts of Salonika, turning north, and followed the well-worn route between the dramatic mountains. He didn't know it, but passing on Dave's left was the city of Pella, the birthplace of Alexander the Great and capital of the ancient Greek kingdom of Macedon. The men took turns peering through the window, but without maps or a compass, they didn't have a clue where the trucks were going.

Dave's mind meandered back to Kansas. Back to the cold, December days of flight training, back to the bitter chill of the cockpit. It was all coming back to him now, and how very wrong he'd been. Kansas wasn't the North Pole. *This* was. Dave was thankful he had decided to pull the woolens over his khakis back at the Triolo Airfield. The fabric quickly waterlogged in the crash, but now it provided a measure of warmth that might save his life. Dave removed half his woolens and wrapped them around the other man freezing by the stove. No one was going to die on his watch.

The next morning, Dave awoke in a cold sweat, feeling miserable. His bones hurt. He couldn't breathe, couldn't move. Every jostle from the bumpy, bombed-out road came up through the floor of the truck into his aching body. During the night, one of the men had surrendered his bunk to Dave, but every time the truck took a hard turn he fell off the bed and landed on one of the fellows below.

The Italian medic had done a fine job quarrying the shrapnel out of his leg with his knife, but the wound still throbbed. In the damp conditions of Pavlos Melas, Dave had developed a cough that ripened into something far more disturbing. Days had passed since he last washed up and he could smell himself. And where was the compass?

The shaking, bouncing trailer amplified Dave's frustrations as the fatigue of the journey found its way to his head. For two weeks, Dave had stayed busy planning an elaborate escape. It kept the wheels of his mind turning. But now, split from his buddies and unable to sleep, he touched his first feeling of hopelessness.

The strafing mission over Salonika had been so carelessly organized and so costly. Dave thought about the lack of planning, the bad intelligence, the decision to put Luttrell in the lead. So many mistakes were made. His mind flashed back to the Greek woman murdered by the Nazis for patching up his flight suit. As God was his witness, he would avenge her death.

Dave could see the refugees used as target practice, the dog chasing the truck. Maybe he should've shot the guards when he had the chance, when his thumb was on the hammer. What about the two squealing Limeys? Those traitors botched every attempt to escape. What cowards. During the second full day of travel, Dave had all the time in the world to turn these memories over in his mind. Through the window he could see the dramatic cliffs flatten out as the trucks penetrated into the heart of Bulgaria. Night eventually fell, and the convoy came to a stop on the side of the road.

The prisoners hadn't eaten for two days, so Dave called out to the sergeant through the window and asked for food. A few minutes later, and to the surprise of every man in the room, the sergeant returned with four large cans and handed them to the prisoners. Their tops had been torn off, and the cans were warm and brimming with stew. The men dug in, using their fingers to pull out whole potatoes and large globs of salty meat, which they shoveled into their cheeks. Chewing was an afterthought.

To Dave, who was accustomed to eating blue and black chunks of rancid horsemeat, the stew tasted absolutely delicious. There were no maggots to nibble on, no grease to sift through. The four men devoured the meal loudly, slurping the can and licking their fingers. They were no longer prisoners of war; they were gods, feasting on the ambrosia of Mount Olympus.

With every passing bite, Dave felt the life flowing back into his body, warming his stomach, replenishing his energy. In a matter of minutes, he'd mopped up the last drop, stretched out on the floor, and drifted easily into his dreams.

CHAPTER 16
RED TAILS

The next morning, the 140-mile journey ended as the trucks arrived at the outskirts of the ancient city of Skopia, in northern Bulgaria. On September 5, 1944, only weeks earlier, Hitler had given orders to establish Independent Macedonia, a puppet state in the Kingdom of Yugoslavia that had been occupied by the Kingdom of Bulgaria.

Four days later, a pro-Communist party within the Bulgarian government committed a coup d'état. The country switched sides, declared war on Germany, made peace with the Soviet Union, and with the backing of the Allied Forces and the Red Army, half a million men from the Bulgarian People's Army were sent to block the German withdrawal from Greece. Now, in mid-October, the Nazis had become personae non gratae, and Dave was among their company.

Hitler's army was hardly the first to be kicked out of Skopia. The city had seen war for thousands of years, changing hands like the baton of a relay race. Before the German occupation in 1941, Skopia had been governed by the Kingdom of Yugoslavia,

and before that the Kingdom of Bulgaria, and before that the
Kingdom of Serbia, and before that by the Albanians. Prior to
1912, it was ruled by the Ottoman Turks for five hundred
years. Before that, it was contested by the Bulgarian and
Byzantine Empires, and before that, on the eve of the first century, Rome had taken it from the Dardanians, who had taken it
from Paeonian tribes, and so forth.

History had repeated itself at Skopia ever since humans first
occupied the area, some seventy centuries earlier. Each century
recounted the same story, asked the same question: To whom
does tomorrow belong?

Hitler's answer to that question took the form of a popular
Nazi motto: "Today Germany belongs to us / And tomorrow
the whole world." The Russians soon launched a fierce rebut-
tal to that question. Dave stepped out of his trailer to witness
Russians dive-bombing the nearby Skopia airdrome. They
had commandeered German Ju 87s, known as "Stukas," and
were strafing the runways with cannon fire. The prisoners
were forced to camp at a safe distance outside the city until
the attack ended.

As the raid unfolded, one of the Germans opened the padlocked door at the back of Dave's trailer and ordered him to
unload its contents. The compartment was filled with aircraft
parts. It suddenly dawned on him. He had been transported
with a Luftwaffe repair outfit.

At the guards' instruction, and with the help of the other
prisoners, Dave piled the pieces into a large heap and set the
stash on fire. In no time, the highly flammable magnesium airplane parts were burning ferociously, and Dave squinted against
the bright white light of the fire.

Dave gathered an armful of parachutes with instructions to
burn them also. But just before he dumped them onto the blazing inferno, a passing thought came into his mind.

I could use this as a handkerchief. Dave tore off a piece of chute and slipped it into his pocket. In seconds, an incensed German major rushed at him, pulled out his P38, and thrust the muzzle into Dave's stomach. He was ready to pull the trigger. Dave froze, not certain whether the German had seen him sheath the piece of cloth.

He had.

Dave quickly removed the torn piece of chute and threw it into the fire. The German withdrew his weapon and stomped off, leaving Dave to bask in the feeling of pure, intoxicating relief. *That man came damn close to killing me.*

For several days, the Nazis filled the sky with flak, aiming at the Stukas that were inflicting damage to the airdrome. The POWs dove for the ditches, covering their heads to avoid the incoming shells and bombs, but the German major who was in charge of the outfit stood in the middle of the highway, shooting up at the planes with his Schmeisser machine pistol.

The explosions stopped late one afternoon, just long enough for the German convoy to transport the prisoners to a jail beside the Monopol state tobacco factory in Skopia. Dave walked into the cell and scanned the room, hoping to recognize any of the inmates. Bro! Milton! Jonesy! Dave made a beeline for his buddies. He also made some new friends. There were five men he didn't know, three Englishmen and two Americans. Dave walked over to greet the Americans, but the dimming light had cast shadows across their faces. No, they weren't shadows, Dave realized. The men must be suntanned. But that didn't seem quite right either. *These men are Negroes!*

For Joe A. Lewis, this wasn't the first time someone had looked with surprise at the color of his skin. In the 1920s and '30s, racial discrimination and law-enforced segregation pervaded the United States. As Europe teetered on the brink of the Second World War in 1938, the U.S. War Department was slow to move away from the prejudiced belief that African Americans were "incapable of operating expensive and complex combat aircraft" due to the color of their skin.

On July 18, 1941, the NAACP filed a lawsuit in federal court to force the War Department to change its policy and accept African American pilot trainees. The very next day, President Roosevelt instructed the U.S. Army Air Forces to conduct an experiment designed to prove the assumption of racial inferiority. In accordance with the prevailing standards of racial segregation, a separate airstrip had to be created. The location was selected in Tuskegee, Alabama.

Joe was among the pilots who received flight training at the Tuskegee Army Air Field. His classroom instruction took place at the nearby Tuskegee Institute, which had been founded by a twenty-five-year-old Booker T. Washington some sixty years earlier on the grounds of a former slave plantation.

As Joe prepared to serve his country, another experiment was taking place at that same Tuskegee Institute. The U.S. Public Health Service wanted to study the ravaging effects of syphilis on the human body, so it deliberately withheld medical treatment from hundreds of impoverished African American sharecroppers in what became known as the Tuskegee Syphilis Experiment. A total of 373 people would lose their lives.

When Joe received his wings on June 30, 1943, he entered a segregated U.S. military. The Civil Rights movement had yet to gain traction. Martin Luther King Jr. was only fourteen years old. The graduate of class 43-F-SE of the Tuskegee Flight School, a second lieutenant from Denver, Colorado, joined the famous 301st Fighter Squadron in Italy—one of only four exclusively African American fighter squadrons to enter into combat in the skies of World War II.

The Nazis were terrified by the thought of Negro pilots and considered the United States unethical for allowing them to fly their planes. But Joe Lewis and his company of Tuskegee airmen would soon leave their marks on Hitler and on history.

On October 6, under the command of the Fifteenth Air Force and flying a P-51 Mustang nicknamed *Precious*, Joe, accompanied by thirteen other "Red Tails" of the 332nd Fighter Group, set out on a strafing mission over Greece in preparation for the Allied troops' invasion. The target was the heavily defended Eleusis Airdrome near Athens.

The War Department considered the mission a success and awarded its flight leader the Distinguished Flying Cross, but the flight wasn't without casualty. At 2:00 P.M., only thirty minutes before Dave was shot down over Salonika Bay one hundred miles to the south, Joe was struck by 88mm antiaircraft flak and crashed in the waters near Athens. He became one of thirty-two Tuskegee airmen shot down and taken prisoner in the Second World War.

Upon his capture, Joe took an awful beating from the Nazis due to the color of his skin. Now, weeks later, he too was shuttled to Bulgaria as a German prisoner of war.

Up until that point in his life, Dave hadn't had many dealings with African Americans, and he wondered how he would get along with the two new fighter pilots. He soon realized, though, that they were far better men in every respect than the two yellow Limeys he'd been managing. Later, Dave would reflect on his first African American friends as being "as good of friends as any man would want in a prison camp."

That night, Dave learned that the prisoners were waiting at the Bulgarian airdrome to be flown out. Flying was the only method of escape, since the Russians had destroyed the northern rails. To the Nazis, captured pilots were valuable enough to warrant the extra effort. Two of the Brits and one of the African American pilots were flown out first. Dave wouldn't see them again for several weeks. The weather closed in as Dave and the others prepared to board their plane, so the flight was called off.

Dave lay on the floor of his cell trying to cope with the high fever that wracked his body. Violent chills caused his teeth to chatter. His forehead throbbed, and his skull felt like it was going to explode. Streams of sweat poured down his chest. For the first time as a POW, but not for the last time, Dave battled a bad case of malaria. He spent the next three days in agony, suffering from excruciating symptoms, willing the disease to leave his body. Joe Lewis stayed by his side. He fed Dave tiny sips of soup and pieces of bread. In between fever spikes, the two pilots hashed out their situation, blew off steam, and shared stories from their personal lives. Talking with Joe steeled Dave against the physical and emotional breakdown nipping at his heels. Above all, it replenished his confidence and reminded him, once again, of the triumph of the American spirit.

CHAPTER 17
THE FLYING
TOOLSHED

Early November 1944, Skopia, Bulgaria

Dave's fever broke on the fourth day, and so did the weather. The skies above Skopia parted handsomely down the middle, and the Germans took full advantage of the opening, loading up the remaining Brits and flying out. Dave never saw them again.

Also waiting to be flown out was a seedy lot of frauleins—the apparent "girlfriends" of the German staff. They meandered around the airfield with greasy hair and threadbare clothes, doing anything to be put on planes. They always took priority over the POWs. Every night, Dave was taken to the airdrome, but just as he would board the evacuation ship, the weather would sock in, the flight would be scrapped, and Dave would return to the confinement of his jail.

The Nazi gun batteries, having offered up most of their ammunition, could no longer prevent the Stukas from punishing the airdrome. With every dive-bombing attack came the promise of freedom if only the POWs could survive a few more days. The food reserves were gone. What little meat the prisoners ate had to be carved out of the charred horse carcasses that littered the airdrome.

Outside the jail, the POWs could hear the cackling of Russian guns belching out in the distance. It was a promising sound, music to Dave's ears, for it meant that the city was falling. Skopia was, indeed, teetering on the edge of liberation. On November 13, after two days of bitter fighting, the Macedonian partisans would drive the Germans out of the city.

But Dave wouldn't be around to see it. Just like at Pavlos Melas, the Nazis shuffled him to another camp at the last minute. It was as if the war refused to let him go. The sad night of his evacuation arrived, and this time it was real. The inclement weather cleared, the Russian strafing ceased, and a Wehrmacht major transported Dave to the airdrome, where he boarded the boxy Ju 52. The passenger plane was filled with German bigwigs and their female companions. Sitting in the cockpit were two German pilots and three young frauleins who'd been tasked with supplying the entertainment. By the time the engines came to life, the airdrome was enveloped with darkness.

It was the first time Dave had flown since his crash. With a top speed of only 165 miles per hour, the Flying Toolshed was a far cry from his sporty P-38 Lightning. Hitler's darling plane was going to be an easy target for the Allies. When the wheels got off the ground, Dave saw a few flashes of artillery burst outside his window. Sitting next to him was a fat, visibly frightened German major who, even in the low light, was unable to hide his terror. The man was clenching a parachute so tightly in his chubby hands that his knuckles had gone white. Dave couldn't help but smile. If the plane went down at their low altitude, no parachute in the world could save him. And Dave would know.

After four grueling hours of hugging the landscape and crossing daringly over Allied positions, the Ju 52 approached a runway at Budapest, Hungary, and started its descent. Just then, one of the Germans shrieked at the top of his voice. The front of the fuselage erupted with bullets. The plane trembled violently as what seemed like a swarm of fiery bees danced around inside the cabin.

A dark shape blitzed by outside the window—a British Beaufighter. The Allied night fighters had the ship in their gunsights and were blasting the hell out of everything. The unarmored Ju 52 was vulnerable to the damaging rounds that easily pierced its thinly corrugated skin.

Bullets zipped from the back of the plane and traveled over the heads of the passengers. One of the frauleins screamed as holes punched through the cockpit. The German major sitting beside Dave doubled over in his seat, his hands wrapped around his stomach. His eyes sank in disbelief as blood spilled out of his belly and saturated the parachute in his lap.

Dave felt the plane lose its confidence as the wings could no longer promise to support the weight of the fuselage. It was a feeling he'd experienced before, that moment of surrender when denial turns to acceptance. The crash was inevitable. Instead of panicking like the rest of the crew, Dave embraced what the plane already knew. He relaxed his muscles, took a few calming breaths, and then closed his eyes. Hopefully, the ship would slow before smacking the ground.

At the last second, the German pilots pitched the nose upwards. There was a tremendous jolt, which elicited loud screams from the terrified passengers. Everything surged forward. Dave felt the floor beneath his feet come alive with violent vibrations as the belly of the plane scraped against the ground, slowed, and eventually relaxed into a stop. The crew fell silent.

Dave opened his eyes to see who was still alive. Somehow,

the German major beside him was still breathing. The cabin itself was riddled with bullet holes, but the plane also proved
resilient. With a few repairs, she might even live to fly another day.

Within minutes, a stream of Nazi GIs spilled into the Ju 52
to evacuate the POWs. Dave was hurried off the plane. He
stepped outside to see dozens of flare pots lining an airstrip.
Several planes burned nearby. The Beaufighters who downed
them were sure to paint a few more swastikas on their planes.

Several weeks earlier, in October 1944, the Red Army was
making progress in its sweep of Hungary to conquer Budapest.
The city was heavily defended by German tanks in the First
Panzer Army and also by Hungary's First Army. By the end of
December, though, the capital was destined to fall. During the
Siege of Budapest, the Soviets would surround the city and
purge the Nazis from the area.

The guards marched Dave and the other shell-shocked men
across the airfield. They passed through the steel doors of an
operations building on the flight line and into a holding cell,
where they would spend the night. Aside from a few bruises,
all of the prisoners had survived and seemed okay. Crashing in
a Flying Toolshed had its advantages.

The next morning, the Germans loaded Dave and the other
POWs into trucks and drove them west, through the wide valleys of Hungary and around Lake Balaton, the largest lake in
central Europe. At only ten feet deep and barely nine miles
wide, the lake stretched forty-eight miles in length—a distance
that would have taken seven minutes for Dave to span flying
full throttle in a P-38 Lightning.

Sixteen hundred years earlier, the Romans had recognized
the medicinal properties of the lake. Before dying of gangrene,
Emperor Galerius ordered the carving of a canal that con-

nected the healing waters of what was then called *lacus Pel-sodis* to the Danube River. During the 1940s, the Germans tapped into the same idea, confiscated the area, and used the oil reserves surrounding the lake to feed their ever-thirsty tanks. They also transformed the lake's southern shore into a luxurious vacation resort, a holiday destination for Nazi officials and their families.

But the leisure was not to last. Within four months, the area would witness the devastating defeat of Hitler's forces as the Red Army crushed Germany's last great offensive of the Second World War at the Battle of Balaton Lake.

When the trucks finally arrived at their destination west of the lake, Dave saw a large train waiting to take him farther. Emblazoned on the side of the train was a long white banner with a thick red cross plastered to its side. It was a Red Cross train!

The Russians allowed one such train to pass the lines each day. But as Dave climbed aboard, he looked around the crowded car to see dozens of SS men masquerading as Red Cross personnel. Apparently, the Germans were hoping to ride to freedom across enemy lines under the protection of their makeshift identification badges, which they disguised by smearing them with ketchup.

The train departed, carrying its motley crew of POWs, Germans, and Red Cross personnel. Dave fell asleep against the window, his bruised and battered body relishing the rest. His sleep was short-lived. Dave awoke to the sound of squealing brakes as the train rolled into the rail yards of Nazi-occupied Vienna. The City of Music was under heavy attack and resounded with a cacophony of screaming sirens and thunderous explosions. The Red Cross train had arrived in the midst of a raid.

Dave was forced to remain on the train as Allied bombers

hammered the railway with their payloads, leaving rubble where buildings once stood. Dave helplessly watched as one by one, the oil cars around his carriage exploded into flames and spilled their liquid fire. Through the window Dave saw the full extent of the fire as the burning bodies of German guards ran by, blazing amid the burning oil. He could see their faces peeling and twisting in agony as the inferno burned their hair and consumed their clothes.

To watch men burn to death was a horrific experience—to see them running, throwing their limbs helplessly, and collapsing to the ground in shock. Dave saw it all unfold only twenty feet beyond his window. He was close enough to hear their high-pitched shrieks, to hear grown men squealing like children as the flames melted skin from bone.

Several hours later, the raid abated. The surviving guards hurried the POWs off the train and ushered them across the yards. By this time, a throng of angry Austrians had stormed the tracks and were following behind, threatening to punish the Germans for destroying their city. Dave watched the mob hurl insults at the Nazis. From the looks on their faces, he didn't doubt for one minute they would make casualties out of the POWs in the process.

Just as they reached the second train, the sirens wailed once more and the skies filled with B-17 Flying Fortresses. More bombs were on their way. The guards had to get out of the open, so they forced the prisoners into an underground bomb shelter.

Dave scurried into the room. When his eyes adjusted to the darkness, he saw large chains hanging from the walls. The only source of light in the room came from outside the barred windows that rose up near the ceiling to meet the street level. Dave soon realized that he hadn't been taken to a bomb shelter. He was standing in an honest-to-God dungeon.

For several days, as the bombing raids refused to let up, Dave would call that dungeon home. The guards kept the POWs away from the windows when the sun was out, but at night the GIs would climb on each other's shoulders and take turns peeking at the city.

Dave's turn finally came. He hoisted himself up and saw a Viennese man riding toward him on a bicycle. The man hopped off, stood on the side of the road near the window, and then emptied the steamy contents of his bladder against a wall.

That was to be Dave's only view of the grand city of Vienna.

After a few days in the dungeon, the air raid ended and the guards herded the prisoners from the bomb shelter into a train. Dave was losing his sense of time, his sense of location.

Under duress, what may have been only five minutes seemed to stretch into the span of five days. It may have taken a few hours for the train to arrive at its destination, but to Dave, who was now delirious with hunger and still recovering from fever, it may as well have been weeks.

But the opposite was also true. In moments of bliss, rare as they were, Dave's sense of time rapidly accelerated. Slurping down that delicious can of stew in Bulgaria took more than a few minutes for Dave to accomplish, but like bullets squeezed out of a 20mm cannon, the adrenaline-fueled experience was over as quickly as it began.

When wracked with a malarial fever or crashing in a Ju 52, a man's entire life can flash before his eyes. All of it came rushing back to Dave—his first flight in the Piper J-3 Cub, the merciless weather of Hatbox Field, the bitter chill of flying in Kansas, the pride on his father's face at the Eagle Pass graduation. When standing on the brink of death, it took mere seconds for him to retrieve and relive those mile-marking moments.

The POWs were muscled out of the train and loaded into blacked-out trucks that careened along for a while and then slowed to pass through something that sounded like a gate. Eventually, the truck halted and Dave felt it go into reverse. The guards pulled him out of the back and shoved him through the door. Everything was muted, gray, and foggy. Dave stumbled down a long hallway that seemed to stretch endlessly, and he was thrown into a small room that was carved out of the corridor. The door slammed behind him.

Dave attempted to adjust his eyes, but in the windowless cell there was not a spark of light. He extended his arms blindly, clawed at everything, sliding his hands against the walls of the room. They felt filthy and left a slimy residue on his fingers. Only the flapper on the cell's door and one tiny vent broke the monotony of the walls. After constructing the room in his mind, Dave laid down on the floor and fell asleep.

The next morning, a sharp odor returned him to his senses. He awoke suddenly and started coughing. Whatever he was inhaling felt toxic to his body, like a million needles piercing his lungs. Dave scrambled to the door of his cell, pulled down the flapper, and pounded. He had to keep breathing, but his body was rejecting the foreign fumes. A few minutes later, a guard appeared outside the door wearing a gas mask.

This happened over and over for what must have been four or five days. Alone in his dark cell, Dave spent his time dreading the next moment when gas would flood in through the vent and fill his lungs. His attempts to plug the vent failed.

Months later, after the war ended, Dave would piece together his experiences in debriefings and compare them to those of his fellow POWs. They concluded that they had likely been held at the concentration camp of Dachau. Dave's cell had been located adjacent to a gas chamber, and each day as the gases

were evacuated, some of the poisonous vapors seeped in through
the vent. For the rest of his life, Dave would live with only 40
percent lung function.

Several weeks earlier, on October 24, as her son awaited transfer to the train yards of Austria, the War Department sent a letter to Dorothy MacArthur.

Dear Mrs. MacArthur:
This letter is to confirm my recent telegram in which
you were regretfully informed that your son, Second
Lieutenant David W. MacArthur, 0714466, Air Corps,
has been reported missing in action over Greece since 6
October 1944.
I know that added distress is caused by failure to receive more information or details. Therefore, I wish to
assure you that at any time additional information is received it will be transmitted to you without delay, and, if
in the meantime no additional information is received, I
will again communicate with you at the expiration of
three months. Also, it is the policy of the Commanding
General of the Army Air Forces upon receipt of the
"Missing Air Crew Report" to convey to you any details
that might be contained in that report.
The term "missing in action" is used only to indicate
that the whereabouts or status of an individual is not immediately known. It is not intended to convey the
impression that the case is closed. I wish to emphasize
that every effort is exerted continuously to clear up the
status of our personnel. Under war conditions this is a
difficult task as you must readily realize. Experience has
shown that many persons reported missing in action are

subsequently reported as prisoners of war, but as this information is furnished by countries with which we are at war, the War Department is helpless to expedite such reports. However, in order to relieve financial worry, Congress has enacted legislation which continues in force the pay, allowances and allotments to dependents of personnel being carried in a missing status.

Permit me to extend to you my heartfelt sympathy during this period of uncertainty.

Sincerely yours,
J. A. Ulio
Major General
The Adjutant General

The letter arrived in Leesville, Louisiana, at 4:30 P.M. on November 2, 1944, and was forwarded to 22 Myrick Street in Allston, Massachusetts, a suburb of Boston and the home from which Dorothy now awaited the safe return of her husband and son.

As Major General Ulio's letter traveled from Washington to Leesville, Vaughn MacArthur began his own journey. On the evening of October 27, the Eighth Armored Division boarded troop trains bound for Camp Kilmer, New Jersey.

Dorothy and her son Charlie escorted Vaughn to the station. Charlie watched as his father stepped up into the train. Just before disappearing, Vaughn turned back to face his family. He smiled and waved farewell as a flash of light caught the metal cross on the chaplain's helmet.

Eleven days later, as much of the United States headed to the polls to cast ballots for either Franklin D. Roosevelt or Thomas E. Dewey, Vaughn and the 2,819 other men of the Thundering

Herd boarded troopships at Pier 45 of the Staten Island Docks and prepared to cross the Atlantic Ocean.

There was no way for Chaplain MacArthur to know that his son had crashed in a Ju 52 Flying Toolshed over Budapest and was sitting alone in a dark cell in Dachau, suffocating from fumes that spilled over from its gas chamber.

CHAPTER 18

"LIEUTENANT DAVID W. MACARTHUR, O-714466"

Unlike the thousands of prisoners who were exterminated at Dachau, Dave left the concentration camp barely breathing but still alive. About five months before the camp was liberated by the U.S. Seventh Army's Forty-fifth Infantry Division, Dave was taken from his cell and loaded onto a train that passed through a series of dark tunnels before arriving at midnight in Frankfurt, Germany.

By the looks of it, the city was all but totally demolished. Stretched out before Dave's eyes were piles of rubble and broken walls, the aftermath of incessant Allied bombing raids. It was November 1944, and for Hitler the writing was on the wall. The Second World War was coming to a dramatic end.

Dave felt a sharp pain in his ankles as he stood to exit, his feet throbbing under the weight of his body. During the train ride, blazing hot heat had been piped up from the engine through coils on the floor of the passenger car. The scorching heat had swelled the prisoners' feet inside their boots.

From the train, Dave painfully limped onto the platform. Nazi guards were standing watch, guarding the prisoners. A German major approached, pointed at the miserable prisoners, and then laughed hysterically. Dave seethed. Already agitated from the journey and barely able to stand, his patience was exceeding the end of its tether.

"What are you laughing about, you crazy son of a bitch?" Dave shouted.

The Germans looked at him curiously at first, then continued laughing. Dave continued his tirade, unconcerned with the repercussions. From the vault of his enraged mind, he conjured up all sorts of curses and other colorful obscenities to launch at the laughing guards. Out of his seared lungs arose the vilest insults the twenty-year-old had ever pieced together, an arsenal of fulminations that he aimed at their smiling faces.

Eventually, the German major lost interest in the raving POW, who had clearly lost his mind, and he walked away. By the time Dave had finished ranting, the transport trucks arrived to take the prisoners to their next stop, Dulag Luft, a German interrogation camp fifty miles north of Frankfurt near the city of Wetzlar, which was designed to hold Air Corps POWs captured in occupied Europe. It would be a temporary stop, an opportunity for Nazi officials to glean valuable information from the pilots before transporting them to their permanent camps.

Established in 1941 to be the primary German Air Force Interrogation Center, the compound of Dulag Luft, or "Auswertestelle West" ("evaluation point West"), was composed of three divisions—an interrogation center located in the town of Oberursel, some fourteen miles northwest of Frankfurt's main train station, a nearby hospital, and a transit camp about thirty-eight miles to the north. Almost every captured pilot was filtered through this interrogation center, which functioned as the Nazi intelligence headquarters for the Western Theater of Operations.

The interrogation center was built to hold two hundred men in solitary confinement. At its peak, though, more than three thousand Allied airmen were processed through each month. At times, as many as five prisoners were crammed into each cell. Four large, white barracks made up the main section of Dulag Luft. Two of the barracks, connected by a passageway and coined "the cooler" by resident POWs, held the two hundred cells.

When he arrived at Dulag Luft, Dave was stripped of his clothes, searched thoroughly from head to toe, inside and out, and then thrown into a solitary cell. There wasn't much in the cell to please the eye. It was twelve feet long, eight feet tall, and five feet wide. There was a blanket on the floor but no cot. There was no window, no lamp, no sanitation arrangements. On the door was a cord. Dave learned that if he pulled it, a flag on the other side of the cell would drop against the door to summon a guard. The constant flapping of flags became a familiar sound to the pilots imprisoned in the barracks.

Over the course of his captivity, Dave had become a shell of himself. The meager rations at Dulag Luft thinned his muscles and reduced his appearance to little more than a skeleton. His once-athletic torso shrunk around his protruding ribs, and all he could dream about, night after night, was food.

All attempts by the Red Cross to provide food for the hungry POWs were rejected by the camp. Dave was given two slices of black bread with jam for breakfast, accompanied by a watery liquid masquerading as either tea or coffee.

The rules at Dulag Luft were strictly enforced. There was to be absolutely no exercise, no smoking, no reading or writing, and no toilet paper. Prior to interrogation, prisoners were usually held in solitary confinement for a total of four or five days. But some POWs, especially the stubborn ones, could be detained in the cooler up to a full month as punishment. One pilot, William N. Schwartz, was held there for forty-five days, far exceeding the Geneva Convention rules.

Conversations of any kind were forbidden in the barracks. There was to be no talking or mingling with other pilots, and even the sound of a whistle could bring severe retribution. Nor were prisoners allowed to meet with chaplains. The only religious activity at Dulag Luft occurred at the hospital when the wounded were forced to attend nightly Bible studies conducted by Hauptman Offerman, the commandant.

Dave spent much of his days fortifying himself with the sparse rations and ignoring the incessant sound of flapping flags. His thoughts drifted forward to his interrogation, to the questions they would ask him, to the answers he would give them. He was well versed in the articles of the 1929 Geneva Convention that required him only to supply his name, rank, and serial number to the interrogators.

Every prisoner of war is required to declare, if he is interrogated on the subject, his true names and rank, or his regimental number.

If he infringes this rule, he exposes himself to a restriction of the privileges accorded to prisoners of his category. No pressure shall be exercised on prisoners to obtain information regarding the situation in their armed forces or their country.

Prisoners who refuse to reply may not be threatened, insulted, or exposed to unpleasantness or disadvantages of any kind whatsoever. If, by reason of his physical or mental condition, a prisoner is incapable of stating his identity, he shall be handed over to the Medical Service.

The most famous prisoner to be interrogated at Dulag Luft was the highly decorated Polish American Francis Stanley Gabreski. During the attack on Pearl Harbor, "Gabby" was among the few pilots to take to the skies to intercept and retaliate against the dive-bombing Japanese Zeros. After being promoted, Gabby went on to shoot down a record-breaking twenty-eight German aircraft, which earned him the leading position as America's top ace in the European theater. Newspapers around the world splashed his name on their headlines. At the time, there was only one other American who could compete with Gabby's accomplishments—Richard Bong, who had downed twenty-eight Japanese planes in the Pacific in his P-38 Lightning.

Just before returning to the United States for a national media tour to encourage citizens to purchase war bonds, Gabby said, "One more mission and I'll go home and marry Kay." On July 20, 1944, during his final flight of the war, Gabby's P-47 Thunderbolt skimmed too low to the ground and crashed. After several days, he was captured and brought to Dulag Luft to be interrogated by the infamous Hans Joachim Scharff, the "Master Interrogator" who coerced his subjects with courtesy, offering them coffee and tea in his personal quarters and even taking them to the cinema. To the Allied POWs who succumbed to the interrogator's tactics and released sensitive material, Scharff was known as "Poker-Face."

Eventually, after days of waiting, the cell door swung open and in walked a guard to whisk Dave away.

The German interrogator at the other end of the table would turn out to be much harsher than Scharff. He fixed his eyes on Dave, studying him, and then he slid a single sheet of paper across the table. The form came from the Red Cross, the interrogator alleged.

Dave scanned the page. No Red Cross agency would ask these kinds of questions. He glanced dubiously at the interrogator, who, in the dim of the light, appeared almost Japanese, which made Dave even less likely to cooperate with the man's questions. Pearl Harbor, after all, was still fresh in Dave's mind. He refused to complete the sham form, slid it back across the table, and simply said, "Lieutenant David W. MacArthur, 0-714466."

The interrogator took exception to the fact that Dave wouldn't sign the Red Cross form. "We will not pass your name on as being a prisoner," he said. "Don't you want your parents, and your buddies, to know you're alive?"

"No. Lieutenant David W. MacArthur, 0-714466."

The questions continued. What is your profession? What is your religion?

"Lieutenant David W. MacArthur, 0-714466."

What is your payment during the war?

"Lieutenant David W. MacArthur, 0-714466."

What is your squadron? What is your group? What is your command?

"Lieutenant David W. MacArthur, 0-714466."

What are the letters, number, and type of your aircraft? Its engines? Power?

"Lieutenant David W. MacArthur, 0-714466."

What was your target when you were shot down? Who were the members of your crew? How many were wounded? How many were killed?

"Lieutenant David W. MacArthur, 0-714466."

The questions continued, one after the next, but Dave refused to answer a single one of them. It was only "Lieutenant David W. MacArthur, 0-714466," and nothing more. Dave was not alone in this response. When the interrogation reports were released after the war, a common notation by the Germans was: "Young, arrogant second lieutenant will not say anything except name, rank, and serial number."

For his refusal to cooperate, Dave was sent back to the cooler. In between interrogations, he sat in the isolation of his solitary cell, only to be summoned back and asked the same series of questions. Each time, Dave gave the same response.

The Germans tried mightily to connect Dave MacArthur with the famed General of the Army, Douglas MacArthur. Was there a relation? Only reluctantly did the interrogators abandon that line of questioning, disappointed that their hunch hadn't been proven fruitful.

At some point during Dave's stay at Dulag Luft, the Germans received a complete dossier on Dave, likely compiled with the help of Italian spies working around Triolo Airfield. The barrage of questions continued, growing in intensity and specificity. Dave assumed the Nazis had acquired some amount of knowledge about his service in the war, and he shifted his tactics, answering with a combination of lies and *I don't knows*. The only truthful information Dave relayed, to the great annoyance of his captors, was his name, rank, and serial number.

Back in his cell, Dave eventually heard something other than the repetitive sound of flags. For several days, he'd listened to the sound, pressing his ear against the door and making out what he thought was a low, pained moan coming from the other side.

One morning, an SS guard who was delivering the daily ration of bread and tea left the door open long enough for Dave to steal a furtive glance across the hall. There was a cell directly across from his own, and its door was also open. Dave looked inside to see an Army Air Corps pilot huddled on the floor, clutching his mangled leg and grimacing.

The guard quickly closed both doors. Dave allowed a few seconds to pass, and when the guard was gone, he called out to the man across the hall.

"Have you got a broken leg?"

He did. It had been badly injured when he'd been shot down.

"Are they doing anything for you?"

"No," the pilot answered, "not till I tell them everything." Like Dave, he had refused to sign the suspicious Red Cross form, and the Germans were withholding medical care until he cooperated.

"What's your name?" Dave asked. His head jerked up at the man's response. Could it be true? "Weren't you in pilot class 44-C at Eagle Pass, Texas?"

"Yes."

"Well, we were classmates there!"

Dave basked in the happenstance and wanted to know all about how a fellow Eagle had managed to end up at Dulag Luft directly across the hall. But there wasn't enough time to satisfy his curiosity. The conversation soon fell silent, and it was the only time Dave ever saw or heard of him.

On November 9, via the German POW mail service known as *Kriegsgefangenenpost*, and in loose accordance with the explicit direction of the Geneva Convention, Dave finally had his first chance since his capture to communicate with the outside world. The interrogators had evidently given up trying to question him. He filled in the required information on a preprinted *Postkarte* addressed to his mother:

I have been taken prisoner of war in Germany. I am in good health— ~~slightly wounded~~

*We will be transported from here to another Camp
within the next few days. Please don't write until I give
new address.*
  Kindest regards

  David W. MacArthur
  2nd Lt 0-714466
  U.S.A.A.C.

Not long after the letter was sent, the Germans shuffled Dave
out of solitary confinement and loaded him onto another train
bound for the transit camp at Wetzlar. A few days later, he would
again find himself on yet another train. Dave didn't know it at
the time, but the Germans had plans for him, plans to make his
next stop his last stop.

Four days after Dave's *Postkarte* embarked on its months-long
journey from Frankfurt to the United States, Dorothy was sent
an update from the Army Air Force's Assistant Chief of Air
Staff who oversaw personnel matters from Washington.

Dear Mrs. MacArthur:
  I am writing you with reference to your son, Second
Lieutenant David W. MacArthur, who was reported by
The Adjutant General as missing in action over Greece
since October 6th.
  Further information has been received which indicates
that Lieutenant MacArthur was the pilot of a P-38
(Lightning) fighter plane which departed from Italy on a
strafing mission over Greece on October 6th. Full details
are not available, but the report indicates that during this
mission at about 2:35 p.m., while strafing an airdrome in
Salonika, Greece, our planes were subjected to enemy
antiaircraft fire and your son's fighter sustained damage.

After coming off the target, crossing the coast, flying out over Salonika Bay, Lieutenant MacArthur's parachute was seen to open. Inasmuch as the intensity of enemy action prevented further observation, the above facts constitute all the information presently obtainable.

There were no other persons in the plane with your son.

Please be assured that a continuous search by land, sea, and air is being made to discover the whereabouts of our missing personnel. As our armies advance over enemy occupied territory, special troops are assigned to this task, and all agencies of the government in every country are constantly sending in details which aid us in bringing additional information to you.

Very sincerely,
E. A. Bradunas
Major, A. G. D.,
Chief, Notification Branch,
Personal Affairs Division
Assistant Chief of Air Staff, Personnel

CHAPTER 19
THE GREAT ESCAPE

Mid-November 1944

From Frankfurt, Dave was transported to his new home, the infamous Luftwaffe-Stammlager or, as it would be remembered to history, Stalag Luft III.

Founded in 1942 and operated by the Luftwaffe, Stalag Luft III was designed to hold only captured Allied pilots and aircrew. After Hitler invaded Poland and spurred the British to join the Second World War, the contest for air superiority led to the construction of the compound where downed RAF pilots could be detained. The camp materialized in the sandy soil beside an already existing POW compound as Russian prisoners carved a rectangular patch of land from the dense fir trees. The site was located in the province of Lower Silesia on the southern edge of the German-occupied town of Sagan, population twenty-five thousand. As the crow flies, the camp was less than one hundred miles southeast of Berlin.

On April 14, 1942, Lieutenant John E. Dunn of the U.S. Navy was shot down by the Luftwaffe and became the first American

POW to join the British in the burgeoning compound. Through mid-June 1944, the number of U.S. Air Force officers in the camp grew rapidly, totaling 3,242 by June 15. After being liberated by the Allies in January 1945, the number of Americans spiked at 6,844. Stalag Luft III became the largest POW camp to detain U.S. officers in Germany.

The camp would first achieve international fame in 1950 by way of *The Great Escape*—a firsthand account of the German POW camp written by Australian Spitfire fighter pilot Paul Brickhill. His book recounts the daring tunnel escape of seventy-six Allied airmen on the night of March 24–25, 1944. Fifty escapees were shot on Hitler's order, twenty-three were returned to the camp, and three made it safely home to the United States.

Thirteen years later, the Mirisch Company released a Hollywood adaptation of Brickhill's bestselling book. Produced by John Sturges and starring Steve McQueen, the blockbuster movie script went through eleven torturous versions by six different writers before its 1963 debut on the silver screen.

"Don't believe everything you see in the movies," Dave would later say. "That damn movie wasn't accurate at all."

After the Great Escape, the Brits were placed in a different compound, and the Americans filled up the North Compound. Eight months later, Dave arrived at Stalag Luft III and joined Brickhill and the other POWs in the same camp where the tunnels "Tom," "Dick," and "Harry" had been dug months earlier. Dave would soon find himself digging his own tunnel beneath the floor of his combine, which was located in the Center Compound.

To keep the prisoners from escaping again, a double barbedwire fence nine feet tall was wrapped around the periphery of the North Compound, where the Great Escape took place. Within the fence was a warning wire, eighteen inches tall, easy to step over but fiercely watched by the guards. Rising fifteen feet above the warning wire was a series of guard towers, or "goon boxes," as the POWs called them. The goon boxes were spaced three hundred feet apart and equipped with machine guns and searchlights. At night, when visibility in the compound diminished and escape became a possibility, a *Hundführer* patrolled the grounds with his Alsatian dog, one trained to go for the throat on command.

In the northern part of the camp was the *Vorlager*, an enclosed area that was separated from the prisoners' combine. It contained several buildings including the German guards' quarters, the prisoners' hospital, a coal storage, and the cooler—a gray concrete building filled with solitary confinement cells. Centered on the perimeter of the Vorlager was the main entrance to the compound, a large gate that was wrapped with barbed wire and heavily guarded.

Dave entered through those gates in mid-November 1944. He would spend two months of his captivity in the adjoining Center Compound, foiling Nazis, digging tunnels, and planning what he believed would be his subterranean breakout.

But digging would not come easy. The sandy earth beneath Dave's compound broke apart easily and was challenging to remove from the tunnels. Due to its darker color, the excess soil was difficult to disguise and had to be shuffled furtively around the barracks via socks hidden within the legs of trousers.

When Stalag Luft III first opened, many escape attempts failed due to shallow tunnels. Hard lessons had been learned. One such tunnel descended only a few feet beneath the ground and one day collapsed under the weight of a horse who stomped its hoof through the shaft.

Like a game of chess between the Germans and the POW escape artists, every move had to be carefully calculated. The guards were Luftwaffe pilots and officers themselves, and they were quick to catch on. To ward off escape attempts, the guards dug trenches around the compound between the warning wire and the barbed-wire fence.

When the prisoners dug deeper, microphones were buried in the ground so the Germans could listen for activity. Soon it became clear that the only way to avoid detection was to dig tunnels at least thirty feet below the ground, which presented a host of problems. Also, to mislead the guards, dummy tunnels had to be dug to serve as decoys.

As Dave adjusted to the life of the combine, he found himself among a resourceful, scrappy lot. The most influential man among the prisoners was Colonel Delmar T. Spivey, the Senior Allied Officer of the Center Compound employed by the Nazis to serve as the liaison between the Germans and their prisoners. In this role, Spivey received better rations and living quarters. But Dave and the other prisoners shared mixed feelings about him. Spivey was respected for his work to improve the living conditions and also for spearheading the clandestine escape committee that existed among the POWs. But at the same time, the Senior Allied Officer worked closely with the enemy to traffic prisoners into the camp. Dave didn't blame him for being two-faced but neither did he take well to the cowardly approach of "go along to get along."

Before Dave arrived, there had been no fewer than one hundred attempts to dig to freedom. By 1944, a network of tunnels, and partial decoy tunnels, had been dug beneath Stalag Luft III. The foundation of the camp, which surely had begun to resemble something like Swiss cheese, was full of wormholes carved out by prisoners eager for escape.

Like Dave, many of the men in the camp were fellow fighter pilots who had flown Spitfires, P-51s, and other top-speed warbirds. Because of their training and ability to make split-second decisions under extreme stress, these men proved valuable to escape attempts and earned reputations for being hot-shots and risk takers. The pilots in Dave's combine were suspicious of him at first, thinking he was rash and possibly even a German spy. It was a typical concern in each of the compounds when new men arrived. The turning point came when one of Dave's buddies from his squadron walked into his combine and vouched for him. In a camp where trust and loyalty were the highest currency, the phrase *Yeah, this guy's okay* went a long way.

"My one aim all of the time was to escape," Dave would later say. That was the aim of every POW, but escaping from Stalag Luft III was a rare feat. The Great Escape had become the stuff of legends. It was a path other POWs were motivated to follow. But since the March escape, the Germans had stepped up their security, checking for tunnels and keeping the camp under constant patrol. Dave never heard of a single case during his captivity where an officer made a successful escape. Several were able to find a way around the barbed-wire fence, but in the end they were always caught.

Organizing an escape attempt from Stalag Luft III was no small task. It took planning, deception, tactics, and skills, and not every prisoner received the same opportunity to bug out. If a man wanted to escape, he had to apply for eligibility before the officers in Organization X, the escape committee founded during the Great Escape.

There was no shortage of skills among the prisoners who belonged to Organization X. No talent was wasted. If men had artistic abilities, they were commissioned to manufacture fake identification documents and passports. Tailors converted uniforms into German civilian clothes. Charming personalities smuggled maps and compasses by developing good rapport with the guards. The resources at Stalag Luft III were scarce, but engineers improvised. They reinforced the walls of the tunnels with wooden bed slats, and they fed oxygen to the men who were digging by inventing makeshift air pumps that operated by hand.

By the time Dave arrived in November, Organization X was running like a finely tuned machine. It had a chain of command, with Spivey at the top of the Center Compound's branch. Beyond the uniforms and the carefully forged documents, the committee taught German to the POWs and required candidates to speak it fluently.

But admission to the committee was not guaranteed. Every newbie had to be vetted. Some prisoners, like Dave's contemporary Paul Brickhill, were deemed unsuited for escape. Brickhill was turned down because he was afraid of confined spaces. That claustrophobia likely saved his life.

Even though Dave was digging a tunnel beneath his combine with the other prisoners, the extent of his proficiency in German still didn't amount to much more than the word *Gesundheit*. Dave was young, and he lacked qualifications—not a compelling presentation for the committee of Organization X. Three times Dave applied for membership. Three times he was rejected with a firm *no*.

If he were going to escape, Dave knew he would have to do it all by himself. A daring and dangerous plan began to crystallize in his mind. The odds of success were stacked against him. He'd likely be recaptured and maybe even shot. But if his timing was spot on, and if luck was on his side, the plan just might work.

If not, at least his attempt would distract the Germans long enough to give someone else a chance to try.

On November 17, an airmailed *Luftpost* left Sagan for *Nord Amerika*, written by a homesick and very hungry *Kriegie* at Stalag Luft III.

Dear Folks,

   Am well and doing fine. Do not worry too much. Red
Cross takes care of the clothing. Send couple pair of
wool socks. In box send half chocolate bars and candy.
No gum or soap. Send ready prepared cake and pancake
flour and Bisquick and cans of spices. Cinnamon, sage,
Bell seasoning, Nutmeg, celery salt. Send baking powder
and concentrated vegetable flakes and concentrated
soups. Don't waste any space. Fill it with candy or food.
Clothing is not important. Put in some vitamin pills and
all the saccharine tablets you can get.

   Will be home soon, I hope. Send package as soon and
fast as possible. Get particulars from Red Cross and post
office. Get weight to the last possible ounce. Mother,
you had better get your cookbook ready when I get back.

   I am a caterpillar now, and not a scratch. Will have
plenty to tell you. Send letters and food. Have you
moved yet? Send gingerbread cake flour. Send cigars or
tobacco coupons they are best for trading.

Dave

One day later, having carved a two-week path through the
angry Atlantic waters, the HMT *Samaria* eased quietly into
the dock at Southampton, England. Though later surpassed
by the *Queen Elizabeth* and *Queen Mary*, the *Samaria* was
the darling of the seas, the finest and most luxurious White
Star ocean liner Cunard had thus far debuted. The four-
hundred-foot-long cruiser began her round-the-world voyages
in 1923 and 1924, and to the surprise of her first passengers,
she was even equipped with an elevator that ferried riders between decks.

   But when the world went to war, the *Samaria*, like everything else in Britain, joined the effort. On September 24, 1940, as Hitler's Luftwaffe blitzed London, she was called to duty and began aiding the efforts of the Children's Overseas Reception Board in evacuating British children to the United States. For the rest of the war, she would serve as a troopship. German intelligence hunted the ship and frequently published false propaganda reports that she had sunk.

The *Samaria* departed from New York City on Election Day, November 7, just as Franklin D. Roosevelt secured his fourth term as president of the United States. On board was the Eighth Armored Division, the Thundering Herd. During the journey across the Atlantic Ocean, the Herders proved unseaworthy. During the first week, the men made frequent dashes to the ship's railing to add the contents of their stomachs to the churning sea below.

The mess hall served food twice a day, but it was prepared in a steam cooker and tasted awful. Those with braver constitutions stood in long lines to be served. Just as long was the queue that snaked away from the ship's store, where tax-free cigarettes could be purchased at five cents a pack.

Lifeboat drills began each day at 9:30 A.M. with the crackling *Now hear this . . .* blasting through the intercom speakers. Next came calisthenics, an hour of German and French language classes, and then free time. The men filled their leisure hours with reading, singing, playing card games, watching films, and ogling one of the nearby ships in their convoy—that ship carried nurses.

It was forbidden to walk on deck after dark, but sleep was difficult to secure. Those who attempted it endured the sweltering heat of crowded rooms like the former dining hall, where hammocks suspended on hooks beneath tables rolled with the motion of the waves. On Saturday, November 11, all personnel stopped at 11:00 A.M. and faced east for a moment of silent prayer in observation of Armistice Day, which had brought the First World War to an end twenty-six years earlier. Three days before the *Samaria* reached her destination, a destroyer escort was alerted to enemy activity near the convoy. Two explosive barrels were rolled overboard. They deployed beneath the water, but there was no evidence that the depth charges found their targets.

For thc troops on board, the coast of England could not have come soon enough. Even the thought of dry land improved the morale of the men. On November 18, the *Samaria* reached land, and the fatigued Herd prepared to step onto British soil. Just before disembarkation, Lieutenant Colonel Grover C. Davis sent a commendation to General Devine regarding Chaplain MacArthur. "Throughout the voyage just completed from New York to United Kingdom," he wrote, "subject officer performed duties as Senior Chaplain in an excellent manner. His untiring energy, attention to details and valuable cooperation contributed materially to the success of the mission."

But Vaughn had another mission in the making. Almost six weeks earlier, his son had been shot down over Greece in his P-38 Lightning. With only the information in the Western Union telegram, Vaughn didn't know whether Dave was dead or alive. All he knew was that he had to find him. Somehow, he had to bring his son home to Dorothy.

On November 24, one week after Dave's first letter left Sagan for the States and his father arrived in England, the starving *Kriegie* made another appeal for food.

Dear Folks,
   Am doing well as can be expected. I have sufficient clothes. We are housed in barracks. I am healthy so do

not worry. Food is one of our problems. I am a hearty
eater. Send rice, spaghetti, beans, powdered eggs, and
bouillon. Anything that we can get—anything that you
can get a lot of in a small space. Weight and space count.
Bisquick and cake mix and spices such as cinnamon and
nutmeg. Send Hersheys chocolate and cocoa. We have
plenty of salt and pepper. Canadian personal parcels are
unlimited. If you know anyone in Canada that could ship
me parcels. Any cost to anyone I will doubly refund on
return.

The point is to get as many parcels to me full of food
and candy. Have it concentrated as much as possible. If
you can get extra stickers in any way. Cans will not come
through and they are too heavy. Send cigars for cig's.
Baking powder and mince meat in cardboard boxes.
Cheeze. Dates, figs, raisins. Half chocolate candy.

Dave

In his two months at the camp, Dave never received a reply.

The hour of Dave's escape from Stalag Luft III finally arrived.
With no help from Organization X, he acquired a pair of wire
clippers, waited until the camp fell asleep, and then crept with
a buddy along the edges of the barracks. With one eye on the
goon boxes, the two men snuck from the Center Compound,
dodging the searchlights that painted a meandering moon on
the ground.

Dave reached the periphery of the camp, stepped over the
warning wire, and made a mad dash to the fence. He checked
the towers. The guards were none the wiser. The wire cutters,
though crude, were up to the task, and Dave frantically snipped

through the barbed wire, layer after layer, clipping each segment as fast as he could. It was taking longer than he'd anticipated.

Somewhere in the distance, Dave could hear German voices shouting. His buddy, ever eager to bolt, was champing at the bit, unable to stand still, urging Dave to hurry. The voices at their backs grew louder, and a few prisoners who were startled by the eruption of noise came outside their barracks to see guards run toward the fence, weapons drawn.

Dave had been spotted. The moon on the ground came rushing at him, and just as his feet went white with illumination, Dave removed a man-sized hole from the fourth layer of fence. For the first time since October, he felt like a free man. Dave pushed his buddy through the fence and then hightailed it into the woods, pumping his legs harder than he had in weeks. All of his training kicked in. For the first time since he'd been shot down, Dave felt the familiar surge of adrenaline, the thrill of speed, the split-second instincts that had saved him so many times in the cockpit.

When the two men had trespassed only a few yards into the woods, the dogs started barking. The spotlight danced around the trees and then came to a rest on Dave's back just as a tremendous force knocked him to the ground. The beast felt heavy on top of him and snarled wildly, thrusting at his neck. In his periphery, Dave could see his buddy trapped on the ground, surrounded by dogs and German officers.

Any hopes for escape were vanishing with every passing second. As Dave shielded his face from the snapping fangs, he could hear guards approaching. One of the Germans restrained the beast while two others jerked Dave to his feet and then marched the two escapees back to the compound with guns at their backs.

The plan had failed miserably, and Dave knew the consequences would be severe. Prisoners had been shot for doing
much less.

As punishment, Dave was thrown into the cooler for twenty-
one days. The dog had roughed him up, and the pains of hunger
continued to dwindle his frame, but as long as there was a fence
holding him in, Dave would find a way to break out.

With limited rations and no human contact in solitary, Dave
slept round the clock and spent most of his confinement turning the botched escape over in his mind. It kept his thoughts
sharp and pulled him out of the depressive rut that so frequently plagued the prisoners of Stalag Luft III's cooler. The
mistakes had been plentiful. The barbed-wire fence took far
too long to cut through. By the time he was in the woods, the
guards had already spotted him.

But freedom had a delicious taste, an addictive and all-
consuming flavor that kept Dave salivating. With or without
the help of Organization X, Dave was determined to savor it
again.

CHAPTER 20
THE CONTINENT

November 19, 1944, Southampton, England

The *Samaria* inched into the Southampton dock as men from the Eighth Armored Division tossed packages of cigarettes over the railing to the Englishmen below.

Vaughn MacArthur stepped off the ship and planted his boots on British soil. The last time he'd seen Dave was at the train station in New Orleans. He never got the chance to send his son his final letter, but it ended with a promise: *I'll be seeing you.*

It was a promise Vaughn was determined to keep.

The voice of a senior British officer came through the ship's PA system. "You've never seen mud like you're about to see at your new training ground." He was right. For months, the Thundering Herd had trained in the hot Louisiana mud. Now it was time to see if those Louisiana Maneuvers had been worth it.

After the men of the Eighth Armored Division disembarked, the Red Cross greeted them with hot coffee and doughnuts. The Herd boarded a British troop train that was bound for Tid-

worth and began the slow and arduous journey. It was a western route that cut deeply into the green heart of England. Along
the way, they passed by Stonehenge, that circular mystery of
old, and then by Salisbury, whose stones and bones had been
the delicacy of pilgrims for seven hundred years.

The two-hour journey wove through a rambling countryside
dotted with quaint English villages. Some troops hopped off
the train and climbed atop thatched roofs for a better view of
the pastoral scenery.

When the Thundering Herd arrived at Tidworth, the skies
opened into a torrential downpour. The march from the train
station to the barracks was as muddy as the British officer had
promised.

Kaiser Wilhelm of Germany designed the Tidworth barracks
in the nineteenth century to provide housing for Queen Victo-
ria's troops. The names of the buildings betrayed their age:
Bhurtpore, Delhi, Jellilibad—names of cities the British Empire once occupied. The barracks were primitive but compared
to the floorless tents that housed the medics, cavalry, artillery,
and infantry battalions, they seemed luxurious.

Vaughn sat down to a hot meal prepared by the Eleventh Armored Division, and then he settled into his quarters. The men
were restless for combat. The Louisiana Mancuvers at Camp
Polk had simulated war, but the men were ready for the real
thing. They were ready to put their training to good use on the
Continent. As the weeks came and went, and as the movement
of tanks and military vehicles heightened the troops' anticipation, there was one solitary question on everyone's minds:
when will we cross the English Channel?

Each day, the men of the Eighth prepared for battle. The
likelihood that some would be captured by the Germans was
high, so classes were taught on interrogation resistance and be-
havioral tactics. Medical detachments spent time visiting near-

by hospitals to familiarize themselves with the kinds of injuries they'd see on the battlefield.

Three artillery battalions relocated to the Royal Artillery Range at Tilshead to test their new weapon, the 105mm howitzer. Three tank battalions were also testing their equipment, the brand-new M-4 Sherman tank, which they took for a spin through the muddy practice course designed to imitate the French terrain.

The Americans were received warmly by the Brits. During downtime, the young soldiers flooded into the pubs of Tidworth to sample the local ales. Chaplain MacArthur, who never touched a drop himself, was tasked with maintaining some semblance of moral order while his troops were guests in a foreign country. He kept an eye on his Herd as they grazed, knowing what was to come.

By mid-December, the Germans hadn't yet managed to add Dave's name to the International Red Cross's list of POWs. Dave's whereabouts were still unknown, but from the Adjutant General's office at the War Department, Brigadier General Robert H. Dunlop sent an update to Dorothy MacArthur.

Dear Mrs. MacArthur:

I am writing you concerning your son, Second Lieutenant David W. MacArthur, who has been reported missing in action over Greece since 6 October 1944.

It is sincerely regretted that up to the present time no further information regarding the status of your son has reached the War Department. A report has now been received, however, which discloses that on 6 October 1944, Lieutenant MacArthur was the pilot of an aircraft which participated in a strafing mission over Salonika

Sedes Airdrome, Greece. The report also states that your son's plane sustained damage from enemy fire, and he was seen to bail out.

Lists of prisoners of War received from the enemy, through the International Red Cross, have been carefully checked, but your son's name has not been found on any of them. The military authorities are utilizing all the means at their disposal to locate our men missing in action, and you may be assured that if any information is received in this office concerning your son, it will be communicated to you without delay.

My sympathy is with you during this long and difficult period of anxiety.

> Sincerely yours,
> Robert H. Dunlop
> Brigadier General,
> Acting The Adjutant General

On the American front lines in Europe, General Patton, the Commanding General of the Third Army, began moving north to smooth out the congested bulge of troops that resulted when the Germans forced their way into Belgium. The Herd spent Christmas Day hosting English orphans, and then they received their marching orders. The situation in Europe had become suddenly urgent. Everything was rushed. Patton needed the Eighth Armored Division quickly, and the Herd scrambled to gather and disperse equipment.

But time had run out. The division would suffer from a shortage of ammunition, and the result would be devastating.

* * *

Back in the United States, on New Year's Eve 1944, a Western Union delivery boy again pedaled his bicycle to the home of Dorothy MacArthur, this time in Boston. Dorothy hoped the telegram would give her information on her missing son's status. But for all her hopes, the bad news came as a surprise.

REPORT JUST RECEIVED THROUGH THE INTER-
NATIONAL RED CROSS STATES THAT YOUR SON
SECOND LIEUTENANT DAVID W MACARTHUR IS
A PRISONER OF WAR OF THE GERMAN GOVERN-
MENT LETTER OF INFORMATION FOLLOWS
FROM PROVOST MARSHAL GENERAL
DUNLOP ACTING THE ADJUTANT GENERAL.

It was January 2, and the first units of the Eighth Armored Division left Tidworth and began their journey to France. They were bound for the French port city of Le Havre and, slightly farther inland, Rouen. But the Herd had a problem. In the scramble to depart, they had brought only 11 percent of their authorized ammunition and were ill-prepared for what lay ahead.

After crossing the English Channel and stepping foot onto French soil, the division got their first taste of the devastating scars the war had carved into the Continent. Both Le Havre and Rouen were utterly destroyed. To make the situation worse, a ferocious snowstorm had dropped the temperatures to below freezing. With little visibility and severely undersupplied, the Herd embarked on the weeklong trek to Rheims, a journey of 175 miles.

When they arrived at Rheims, a German reconnaissance plane flew overhead. The plane, nicknamed "Bedcheck Charley" for its propensity for flying in the evenings over the division's bivouac area, was an ominous sign of things to come.

The next day, the Eighth Armored Division received its assignment. They joined General Patton's Third Army and moved to Pont-à-Mousson, France, in preparation for an impending German counterattack at Metz. The men had to disguise themselves for the mission. They covered the markings on their vehicles, removed the patches on their uniforms, and from that moment on the Thundering Herd adopted a new code name for the division: *Tornado*.

General Devine received orders to prepare the first units for battle indoctrination. The long-awaited combat was about to begin.

Seventeen days into the Tornado's wake, on January 19, Vaughn MacArthur penned a letter to his bride back home. He listed his whereabouts as simply "Somewhere in France."

Dearest,

I am writing this Air mail so that I can enclose the two letters I received about Dave and you can file them away for future reference. In years to come they will fit in well in the scrapbook.

Yesterday we received our PX Ration and we got 6 packages of cigarettes, 5 bars of candy, 4 packages of gum, a toothbrush, a package of shaving cream, all for 37 cents. Then this morning we got our QM ration which is free which included 6 packages of cigarettes and 5 bars of candy and some razor blades so right now we have cigarettes and candy floating around everywhere.

It rained during the night and as a result the roads are a glare of ice but the cars and trucks are running just the same. It may get colder tonight and it may warm up. It has been fairly cold here for the last two weeks but it

ought to be beautiful here this summer and spring and by
the time fall comes around I hope that we shall be on our
way home.

The news that we get here is that the Russians are
going to town but they have not yet crossed over the
German border. We haven't made too much progress in
the Reich yet either but I hope that they are waiting until
the all out push comes.

While passing thru France I was able to buy some
postcards which I am going to send. Most of our moves
were late in the afternoon and during the night so I wasn't
able to stop in any towns where we could purchase postcards. Some of the men are taking pictures of different
places and things and I have asked them to make two
prints of each picture for me but as yet they have been
unable to get the prints developed so that doesn't help
me any.

I received a letter from Bob Horton the other day in
which he said that Charles Kiehle went to his church and
told about Dave being missing. I shall have to write to
Bob and tell him the latest news. While I am miles apart
from him in some of his views I like Bob and admire his
stand for he has very strong convictions along that line.

The other day I received some of my letters to Dave.
They were returned and I opened one which had the clipping of his chapel. I am sending it back to you just in
case when he gets back it might have been his chapel
and the one we picked out might have been him.
Wouldn't that be a coincidence if that was the case?

I hear that Dan Fielder is much closer to us than I
thought at first. He has moved so if I get a chance I must
venture up his way because I broke one of the teeth off
my upper partial which I must get fixed right away and I

have to go north from where I am to get it done. That
would take me close to where he is. That would be great
to see Dan again.

After looking at the weather I think I ought to go forward and see about the services for Sunday. We tried to
use the Catholic Church here for the Protestant and
Jewish services but it was no go so I shall have to go
over to the Hotel in the town and see if I can use that
for the services. The trouble we have here is over the
confessions. The French priests cannot understand
English so we have to have our chaplains make the
rounds for the men.

Well dear I had better hang up and get the stuff in the
mail and then get my business attended to and eat early
chow and get on my way. I love you dear and miss you
and long to be back with the family.

Love
Vaughn

To get their feet wet with war, the Tornado division had been
scheduled to undertake a forty-eight-hour battle, but a harsher
indoctrination awaited. For the next seven days, the men engaged in nonstop activity with the Nazis. Twenty-three men
were killed in one battalion alone, and another 268 were
wounded. By the end of the week, after suffering terrific losses,
it was agreed that the Eighth Armored Division "had been up
against some of the best of the Wehrmacht troops on the
Siegfried Line."

General Devine praised Vaughn's men. With regard to one
particular unit, he said, "Never have soldiers behaved better in
their first contact with enemy troops than did those of Combat

Command A during their attack on the Siegfried Line Switch Position."

Chaplain MacArthur didn't flinch in the fight. During that first week of battle, he spent his entire time tending to the needs of his beleaguered Herd. He prayed with the wounded, moved casualties to safety, boosted the morale of his men, and administered last rites to the dying Catholic soldiers. Soon, he had earned the respect of nearly every man in his division. For the rest of the war, if a soldier needed Chaplain MacArthur, he knew where to find him—on the front lines in the hottest fire of battle.

More than five hundred miles away, Dave shivered in the camp of Stalag Luft III. Vaughn's son was about to embark on his own journey. At 9:00 P.M. on January 27, Colonel Spivey burst into Dave's barracks and roused the sleepy POWs from their sacks. "Get ready," he ordered. "Everyone will be marching out in half an hour."

CHAPTER 21
DEATH MARCH

January 27, 1945, Stalag Luft III, Sagan

For weeks, the Germans had kept a close eye on the Russian armies as they forced their way through occupied Europe. Rumors of the advancing Red Army consumed the prisoners of Stalag Luft III.

Some two hundred miles to the southeast of Sagan, the Russians entered the Auschwitz camp complex and liberated more than seven thousand prisoners. Most of the emaciated POWs were gravely ill and close to death. The SS had conducted a mass extermination at Auschwitz in the weeks prior to its liberation, evacuating roughly sixty thousand men and women in a series of death marches westward. More than fifteen thousand prisoners would die on those marches.

At least 1.3 million prisoners—mostly Jews—were sent to Auschwitz between 1940 and 1945. More than 1.1 million were killed.

With the Russian troops on the cusp of the German POW camps, Dave would now begin his own death march west.

* * *

The prisoners of Stalag Luft III were given thirty minutes to prepare to leave, and they scrambled to gather their scanty belongings. Some men fashioned knapsacks out of their thermal underwear, knotting their trousers to carry as much food as possible in the legs and seat.

In preparation for the march, Dave grabbed two straw palliasses and tore their covers into long strips. He broke his bunk into planks and used the fabric strips to tie together his makeshift sled.

Sagan was in the middle of a fierce snowstorm, one of the coldest in recent memory. The men donned their overcoats, grabbed their blankets, and loaded their knapsacks and sleds with as much food from the Red Cross parcels as they could find. They abandoned letters, books, and camp records, and waited for the order to evacuate.

Over the next two hours, every prisoner of Stalag Luft III, save some two hundred who were deemed too weak to walk, was lined up. Flanked by German guards carrying machine pistols and rifles, they marched out in three columns to face the piercing cold.

The airmen of the South Compound were the first to leave. Between 1:00 A.M. and 3:15 A.M., the North and East Compounds followed suit, with Dave's Center Compound right on their heels. Dave, in one final act of defiance, set fire to his barracks just before his combine stepped out into the snow.

The Germans gave the prisoners a sobering warning. If anyone fell out of line for any reason, he'd be shot. One of the men standing near Dave—a pilot who'd been all talk in the camp—lost his bravado and morphed into a "sniveling dog." Frustrated murmurs flowed among the columns as the other prisoners speculated about how much of a liability the man would be.

The columns inched forward. Dave's lungs, damaged from the poisonous fumes at Dachau and now inflamed with bronchitis and pneumonia, ached as the pilgrimage of men passed through the camp's main gate.

They came to an enormous building at the entrance of Stalag Luft III, where personnel were handing out Red Cross parcels.

For months, the POWs had been forced to survive on reduced rations. When Dave realized that the guards had been stockpiling the Red Cross rations instead of distributing them among the prisoners, anger replaced the fear that had been festering in his stomach.

Keep them hungry and they won't give you any trouble, he surmised. The prisoners broke from the columns and began snatching up the rations. Most could carry only one Red Cross parcel, but with his sled, Dave piled on four.

Civilians had gathered just outside the fence, waiting for the camp's evacuation so they could swarm in and loot the barracks. As his column marched out, Dave turned to steal a last look at what had been his home for the previous three months. His barracks, and at least half a dozen others, were burning brightly in the night as the fierce wind carried embers across the various compounds.

Onward they marched, and at first the men felt almost giddy to have escaped the camp. But the feeling soon gave way to exhaustion as the bitter cold tested each man's stamina and spirit. Dave teamed up with a prisoner who was struggling to walk. He eased the man's burden by carrying his pack for him, and helped him as he stumbled along the way.

The prisoners marched throughout the night, stopping only once an hour for a ten-minute break. When the guards grew tired, they forced the prisoners to carry their packs. Dave, with his sled, became a natural choice to haul the extra loads.

* * *

Eventually, some of the men died from exposure. The Germans were terrified of the Russians they captured and forced them to march barefoot in the snow until their feet became blackened, frozen stumps. But the POWs knew the Red Army must be close, and they did everything they could to stall their own procession.

On the second day, a senior officer from the Center Compound suggested that the prisoners combine all of their food and distribute it more evenly. Dave was forced to surrender his four Red Cross parcels, but by the next morning, the officer changed his mind and ordered the food to be returned to its original owners. Dave never saw his three extra parcels again.

Sometimes, the prisoners slept in unheated barns, which did nothing to halt the raging snowstorm. During the day they bartered furtively with the curious civilians who flanked the road, exchanging cigarettes for food. What had started as an orderly processional of three columns deteriorated into a disorganized mob of men struggling to stay alive, struggling just to take the next step.

At night, as the exhausted men settled down to rest, one of the prisoners would inevitably have a breakdown. These men were usually dangerous. Once, Dave saw a man go berserk and charge a guard, knock him down, and take his gun. The other prisoners hurried to intervene and were finally able to quiet the disturbed pilot. He seemed to be alright after that. To Dave, it seemed mainly just a matter of releasing the pent-up tension that mounted as the days went by.

One night, the prisoners came upon an old church along their route. Thousands of them poured into a sanctuary that had been designed to seat only two hundred parishioners. Men crammed into its quarters, spilling onto and underneath the pews. Bodies were strewn down the central aisle in a tangle of overlapped limbs. Men chose to stand shoulder to shoulder against the walls all night rather than endure the blizzard outside.

Once the church was full, the German guards locked the doors. For those unable to squeeze in, there was another option—the church's graveyard outside. To escape the freezing wind, men broke into vaults, opened the coffins, and slept with the dead in hopes of not becoming one of them.

At dawn, the tombs were emptied of their living; two men inside the sanctuary had died. Prodded by the guards, the POWs kept marching. After several days, Dave could no longer stand up straight. Every time the column stopped, he fell over in the snow and laid there, unable to move. When it was again time to march, stronger prisoners pulled those who had collapsed to their feet once more.

The POWs weren't the only ones to suffer. The harsh conditions and meager rations affected the guards and their prisoners alike. During one break, a guard sat down on the side of the road and fell asleep. Within a few minutes, he had rolled over on the ground. When one of the other guards tried to wake him, he found the man was dead. Every night, some of the guards deserted.

One day, the prisoners began to sing to keep their spirits up. As they marched through a town, the streets were lined with armed *Volkssturm*. The "people's army" clapped and waved as the pilots sang the Army Air Corps song at the top of their voices. The pilots tried to harmonize and were delighted when a rotund, good-humored Nazi guard joined in and imitated a tuba, his large belly swaying back and forth as his lips burst forth their staccato accompaniment.

He's pretty good! Dave thought. That night, the guard deserted.

The next afternoon, as the prisoners marched, one of the guards thought he saw a U.S. plane swooping down for a strafe.

Certain they were about to be struck, the nearby prisoners waved everyone off the road and took cover in the woods.

The German guards in the front and rear were beyond earshot of the frightened men in the middle and assumed the prisoners had executed a mass escape attempt. The guards opened fire on the group. They killed three of their own men and wounded several Americans who had taken cover in the deep ditches along the road. In the ensuing chaos, several prisoners decided to seize the opportunity and make the false escape attempt a true one. They took off into the woods—the first of many escapes along the winter march.

Dave watched man after man try to escape. After he'd nearly consumed his lone Red Cross parcel, he decided to give it a try. His frustration over his lost parcels—and the dog-eat-dog attitude of the men who had commandeered the rations—emboldened Dave to make a run for it.

That evening, as the Center Compound parked for the night, and with the last of his food stuffed into his pockets, Dave wandered off into the woods under the guise of taking a leak. Dave walked all night through the silent, deserted country. Sometime in the wee hours, he stumbled upon two of the fellows who had escaped two days earlier. Unbeknownst to the men, they had inadvertently walked parallel to the same path of the march.

The three men took off through the woods in another direction, eventually coming to a small farming community run by displaced French forced laborers. A family put Dave and the two men up in their own beds, fed them abundantly, and let them listen to the radio.

For the first time in months, Dave enjoyed the luxury of heated water. He removed his filthy prison clothes and dipped his thin body into an actual bathtub. Once he was washed, the family provided Dave with clean clothes to wear. For two glorious days, the three men found refuge in the warmth and comfort of the farmers' hospitality. On the third day, though, ten SS troops screeched to a halt in their trucks outside the small cottage. They barreled inside, grabbed the three Americans, dragged them out of the home, and beat them mercilessly.

Dave would never find out who had tipped off the Germans, and God only knew what price the French family paid in exchange for harboring the POW escapees.

Dave and the two men in his company rejoined the death march at the back of the line with the men from the Center Compound of Stalag Luft III. Still aching from their beatings, they struggled along for two more days. Dave came down with such a dreadful case of dysentery that what little nourishment remained in his system made its way out from both ends of his body.

After marching thirty miles or so, the prisoners arrived in Muskau, a quaint thirteenth-century town located on what is now the Polish-German border. With frostbite nipping at his fingers and toes, Dave was edging along the brink of utter exhaustion. The prisoners from the South Compound had made the journey from Sagan to Muskau in twenty-seven hours, stopping only once for a four-hour sleep. Dave's Center Compound was days behind.

When Dave arrived in Muskau, he was marched into a porcelain factory filled with giant clay urns fresh from the kiln. Dave sized up his surroundings and noticed that some of the urns were large enough to hold a grown man. He found an enormous one, crawled inside it, and drifted to sleep. The urn, still warm from the firing, gave Dave his first good night's sleep in days. The next morning, the men were marched back into the falling snow. Men too weak or sick to continue were left behind. Their exodus had covered thirty-five miles, but there were many still to go.

* * *

From Muskau, Dave and the men from the Center Compound followed in the footsteps of the South Compound. They set out on the sixteen-mile journey due west and arrived at a military school in Spremberg. There, the prisoners rested for part of one day before the guards marched them out again that night. For the first time, though, the guards chose not to lead the columns to the road. Instead, the men were marched down to a railroad yard and were loaded into French World War I–era boxcars. Emblazoned with a large *40/8* on the side, the "Forty-and-eights" were designed to hold forty men or eight horses. Dave and fifty-one other men were loaded into the last freight car. After they settled in, the guards loaded forty more men into their car. Nearly a hundred bodies crammed into a boxcar built to hold half as many.

The South Compound made the journey from Spremberg south to Moosburg in two full days and nights, but as the approaching Allied forces carried out nightly bombing raids, it took the men of the Center Compound much longer. For five days, Dave sat on his heels, squatting against the side of the car with only the balls of his feet on the floor. For nearly a week, he wasn't able to move.

For the first few nights, when the boxcars came to a halt as the Allies began dropping bombs, the German guards locked the doors of the freight cars and absconded to air-raid shelters. By mistake, they locked a few of their own guards inside with the POWs. The panicked guards shot the lock off the door and escaped.

In a bombing raid on the third night, the man squatting beside Dave—one of his closest buddies—went mad. He shouted and flailed his arms. His frantic eyes searched in vain for a way out of the freight car. Dave held him down, and for two days he kept a close eye on his friend. For Dave, agitation and fear, mixed with the tight quarters, now made sleep impossible.

On the fourth night, when the piercing whistles of bombers overhead approached the slow-moving freight cars, the train engineer panicked and accelerated to breakneck speed. Dave's heart raced as the man at his side babbled incoherently. The train pitched through the dark, whipping Dave's boxcar from side to side, and nearly derailed the train. The hundred men in Dave's car moaned and cursed. Sobs and prayers filled the suffocating space. When the train finally slowed to a stop, the men breathed a collective sigh of relief. Two prisoners looked at the guard next to them and issued an ultimatum: if he didn't open the door and let them out for at least a minute, they would break his neck and throw his body from the train. Seconds later, the battle-weary prisoners spilled from the boxcar into the cold night air, surprised to be alive.

On February 7, eleven days after Dave had taken his first step into the snowstorm in Sagan, he arrived in Moosburg—his home for the next three months. The name of the camp was Stalag VII-A. One week earlier, on February 1, Dorothy MacArthur had attempted to send a letter via airmail to Prisoner of War #8754 in Stalag Luft III, hoping to inject some warm memories of home into her son's imprisonment.

Dear David:

I am down at Myrick Street taking care of the children while Art and Bernice are out for the evening. I agreed to take care of them one night a week. I am quite busy these days with my school work. I have some notes to type but I am sort of tired. I don't know what to tell you that would be of special interest. Charlie is working at the Bakery. Tonight is choir rehearsal. I wonder if you received my letter with his picture. He took some flash pictures at Christmas time but they didn't come out very well. I haven't received any word from you yet. I was going to put in some writing paper in your box but they told me you would have some. I wonder if you have ink for your pen. I sent six more cartons of cigarettes because I knew they would be welcome if they are as scarce there as they are here. I think Art and Bernice have gone to see Hazel and Jack. I am having the whole Mullen crowd for dinner a week from Sunday. It is a tight squeeze in our small place. Last Sunday we went to Edith's. The next time we will go to Ruth Crosman Bailey's. Little Kenny is growing fast. He looks like the rest of the family but he has a boyish look. He is quite a novelty in that bunch of girls. Grandma says to give you her love and she hasn't forgotten the promise you made that she is going to be fixed up in her old age. Bernice and I are taking swimming lessons at the Y.W.C.A. We tried to dive last night. She went down too deep and I didn't go down deep enough. You know the way I dive. Perhaps I will learn to do it right. I am kind of an old girl to be learning but they say we are never too old to learn. I am thinking about you my Davy and hope the World will be at peace before long and all the Mothers all over the world will have their boys home again. I love you.

Mother

Little did she know that Dave was no longer a prisoner of Stalag Luft III. Months later, Dorothy's letter would be returned to her. It was marked by the War Department *Undeliverable as Addressed*. Dave was already on the move.

CHAPTER 22
STALAG VII-A

The Eighth Armored Division was also on the move.

On January 13, the Herd moved to Thiescourt, France, just west of Noyon, where they remained for the next three weeks. Always mindful of the morale of the men, General Devine and Colonel Charles G. Dodge arranged for the Division Special Services Officer, Major Henry Rosenberg, to scrounge up some entertainment. No one quite understood how he managed to pull it off, but a dance was arranged, complete with fifteen young ladies who were willing to attend. The troops were delighted to see General Devine and Chaplain MacArthur kick off the festivities by finding partners and heading out onto the makeshift dance floor.

Just as the dance drew to a close, and as the young ladies filed into 6x6 trucks for the ride home, an ominous sound echoed through the evening air. Vaughn looked up to see squadrons of heavy bombers filling the sky. The town erupted with fire and explosions as five-hundred-pound bombs pummeled the area. The men manned their posts, aimed their wea-

pons at the sky, and unleashed a mountain of artillery at anything flying low enough to hit. One plane was hit, and its crew
bailed out. With revolvers drawn, men who had been dancing
only moments before sped off toward the downed aircraft to
capture the survivors. The time for grazing was over. The Herd
was fully at war.

On Valentine's Day 1945, as the Eighth Armored Division was
headquartered in Holland, Vaughn received a welcomed V-mail
message from his wife.

Addressed *To my Valentine* and penned within a hand-sketched
broken heart, Dorothy composed a hopeful poem for her husband:

> *My heart is broken as you see*
> *Because my love is far from me*
> *I hope he'll soon come home to stay*
> *And I will sing Hip! Hip! Hooray!*
>
> *Dorothy*

As his father pushed through Europe, dodging bombs and en-
gaging in skirmishes with the Nazi forces, Dave found himself
in one of the most miserable places he'd ever been—Stalag
VII-A. The compound was located just outside the city of
Moosburg, about twenty miles from Munich. It had been built
as a prisoner-of-war camp to hold noncommissioned U.S.
Army Air Corps officers, then it served as a transit camp for
ground troops. By the time Dave arrived, the camp had become the final stop for Allied officers shuttled from other
camps across Germany.

Stalag VII-A was composed of three compounds, and the
entire camp was surrounded by layers of barbed-wire fence.
Beyond the fence were hills, except for a small field on the

eastern side. The camp was originally built to hold ten thousand prisoners and had plenty of barracks for its normal capacity, but the compounds were woefully inadequate for housing the thousands of additional prisoners who had marched to Moosburg in January and February 1945. By the end of the war, every square foot of the camp was occupied by men sleeping in tents.

The Red Cross provided food and clothing for the prisoners, but after the arrivals of POWs from the winter march, there was never enough. Camp doctors ran out of medical supplies. Lice and fleas ran rampant. The latrine pits were constantly in need of emptying. If a prisoner got a shower every two weeks, he was considered lucky. Hot water became only a memory. In the damp and cold, the prisoners stayed sick, and the entire enclosure was in a continual state of chaos as the German guards tried to instill order among the thousands of men.

Surprisingly, though, the morale at the camp was usually fairly high. The winter influx of spirited Allied officers from surrounding camps did much to raise the disposition of the prisoners. One night, Dave struck up a conversation with some of his old P-38 pilot buddies from Triolo in Italy. Each of them had taken a different route to Moosburg, but now it seemed as if every POW in the entire country had ended up together at Stalag VII-A.

"I have no idea if my father is overseas yet," Dave told them, "but if he's anywhere in Europe, I know he'll come down and get me."

Not long after he arrived at Stalag VII-A, Dave began planning one final escape. He had first tasted freedom at Stalag Luft III back in Sagan, but this time, after thinking, plotting, planning, and scheming, he hatched a foolproof plan. This time, he'd do it alone. In his mind, Dave visualized the escape. He would secure wire cutters, cut through the fence, and catch one of the trains that passed through the area.

Late one evening, while the men in his overcrowded tent were fast asleep, Dave snuck out. He'd managed to fashion a pair of wire cutters and made a run for the fence at the edge of the camp. One layer after the next, Dave slashed through the wires until an opening appeared large enough for him to slip his body through.

Dave raced through the woods, fully expecting to be chased by barking dogs. But unbelievably, the night was quiet. He'd really done it this time.

Dave ran until his weakened lungs heaved and burned. He indulged himself in a quick respite, no more than a minute or two, before once again picking up his pace and scrambling through the woods and hills.

He eventually stumbled upon a train station where he hid out of sight until he could make sure no one was watching. As one of the trains was pulling away, he mustered one last bit of energy, one last sprint. Hopefully, it would be enough. Dave bolted for one of the freight cars that still sat motionless on the tracks. He ducked, crawled under it, and held himself beneath the crossbars.

The car began to move. For three days, Dave held onto those crossbars for dear life. Each time the train stopped, he received a few precious minutes of relief when he could lower himself onto the track, relax his muscles, and try to sleep. The bitter cold stiffened his hands, and with little more than his threadbare prison garb to wear, Dave nearly froze to death.

Three days into his escape, and just before he reached the Swiss border, the French boxcar slowed at a checkpoint and German dogs sniffed him out. Their angry barks alerted the nearby guards. The two men yanked Dave out from under the

train, jabbing the butts of their guns into his ribs, and back to
Moosburg he went. Back to the cooler for twenty days of
hunger and isolation.

For three months, Dave's entire world was Stalag VII-A. After
his twenty-day stint in solitary, he rejoined his crowded tent
and tried to pass the time as best he could. He dreamed of his
mother's cooking, the feel of a warm bed, the G.I. Bill waiting
for him at home. He obsessed over his crash in the P-38 Lightning. Should he have tried to feather his bum engine before
bailing out? It was a thought he rotated in his mind a thousand
times.

As Stalag VII-A began to feel unbearably claustrophobic,
Dave found himself fantasizing about his tent back at Triolo.
By any estimation, the tent was an engineering marvel, a
masterpiece of ingenuity. With the help of several of his tent-
mates, each who helped secure building supplies from around
the airfield, Dave had created this Italian home.

Every detail of the tent stuck out in his mind. It had a floor
of stolen bricks on which perched a flimsy structure of belly
tank crates. Over the crates, Dave and his buddies slung two
pyramidal tents with space in between to provide insulation.
They even fashioned a door and two windows made of crates
and cellophane that had been wrapped around the protective
gas covers.

Just outside the tent was a small rack for their helmets,
which doubled as washbasins. Beside the rack was a belly tank
mounted on supports that Dave had managed to drop on purpose while taxiing on the ramp in the P-38. A line ran from one
tank to the basins. Another filtered into a fifty-gallon drum,
which one of the crew chiefs had cut down. On the drum was a
steel plate connected to hundred-octane gas. When the boys
threw in a match, it made for a perfectly hot stove. The only

trouble, of course, was that the stove often blew up. Many a good tent had been lost that way.

During the freezing nights at Stalag VII-A, Dave remembered the sacks in his tent, the GI canvas cots on which he'd layered air mattresses, sleeping bags, and a couple blankets. What he wouldn't give to be back there in Triolo, to be wrapped in those blankets, to enjoy even a single night at that Italian airbase.

Over the course of his months in the compound, Dave's mind grew restless. Each night, as he struggled to sleep, the kaleidoscope of his thoughts shifted, and shards of older memories came into view.

He saw his father talking on the telephone, convincing Clemson A&M to release him from the Infantry. He heard his father mention that the Aviation Cadet Selection Board was coming to Greenville, and that he should consider applying. Hundreds of letters scrolled by, letters Dave and his parents exchanged during his two years of flight training.

Then came graduation day. Dave felt his father's hands pin the wings to his chest and heard the graduation speech clear as crystal. Success, the chaplain had said, can only be achieved when everyone works together for victory.

Suddenly, Dave was in his P-38 Lightning again. He felt his body go weightless, then heavy, his thumb on the trigger. Something exploded. He smelled gasoline sloshing at his feet. He felt the flames against his face and the icy waters of Salonika Bay.

All the horrors of Pavlos Melas rushed back—the morning executions, the ditch full of corpses. There was a young girl crying. Dave saw the Nazi shoot her mother and kick the child into the ditch. The white powder filled the child's mouth and eyes. Another woman appeared, a Greek woman, slain by the

Nazis for sewing up Dave's flight suit. Never had such kindness and cruelty come so close together.

Then came the interrogations. Dave heard himself repeat the mantra "Lieutenant David W. MacArthur, 0-714466." He saw himself running into the woods, chased by dogs, trying to escape. His feet went numb in the sixty-mile death march to Moosburg.

Eventually, Dave's memories would fade. And every night, before he succumbed to sleep, in those lonely moments when his home felt farthest away, Dave said a prayer for his mother and then escaped into his dreams.

CHAPTER 23
THE SEARCH

For the chaplains of World War II, conducting religious services in combat was a test of adaptability and ingenuity. With a limited number of chaplains in his purview, Vaughn often found himself functioning as a Catholic priest, a Jewish rabbi, and a Protestant minister. He did his job so well that some troops of the Eighth Armored Division had a hard time pinpointing his exact religion, but they could always find him because of the silver Christian cross soldered to the front of his M-1 steel helmet.

On one Sunday, a small group of soldiers from the headquarters of the 130th Armored Ordnance Battalion celebrated Mass in an open field in Germany with "Father MacArthur." The chaplain's assistant was a soldier named Charles Terrell, a former soloist with the Rockettes in New York City, who led the group in singing a few hymns accompanied by his portable organ.

Chaplain MacArthur used the hood of his jeep as an altar for communion.

Vaughn also had another mission, a personal mission. He requested to be stationed on the division's front lines, and as the Thundering Herd blazed its trail through Europe, he searched for his son. Vaughn didn't know if Dave was even alive, but if he was, he would find him and set him free.

From Valentine's Day until the middle of April, the Eighth Armored Division passed through Nazi-controlled cities, towns, and villages. Their tanks rumbled menacingly over the terrain. Having sojourned from England to France, and then from Holland to Germany, the Herd was now well seasoned. They enjoyed occasional times of rest and fun, but every man kept a steely eye on the task at hand, on the next mission, on the next battle.

They trudged through the German Rhineland, headquartering in towns whose names most of them couldn't even pronounce. From Hückelhoven on the German border they traveled north to Lobberich, where they spent the month of March. From there they went east toward Zweckel. Near the end of their two-week stay in Beckum, Germany, some men of the Eighth Armored Division were assigned a new task.

In the village of Langenstein, a medic learned of a nearby concentration camp. He ventured out to find it. When he entered the gated compound perched atop a hill, the man knew he would never forget the horrors he saw. He'd found Langenstein-Zwieberg, one of the dozens of satellite subcamps of the concentration camp of Buchenwald. The medic went back to his headquarters to report his findings, and within hours, he returned with enough men from the Herd to liberate the camp.

The compound was strangely silent as they walked through the gate and down the barracks. Packed into the overcrowded bunks were emaciated bodies whose hollow eyes followed the Americans. Approximately one thousand prisoners were still alive when the Herd took control of Langenstein-Zwieberg.

But the average prisoner weighed about sixty pounds, and by the end of the next week, only one hundred still survived.

The experience had a profound effect on many men in Vaughn's division. He did his best to comfort and counsel them, but he himself had also been affected. Hitler had just made the war much more personal.

In April 1945, Vaughn was one step closer to finding his son. Technical Sergeant Salomon Braus* was a frontline interrogator with the Eighth Armored Division's MII Team 436-G. A German Jew, he had immigrated to the United States nine years earlier after enduring severe beatings by Hitler's SS. In 1941, five years after leaving Germany, Salomon joined the U.S. Army. After completing his basic training with the cavalry at Fort Riley, Kansas, he was ordered to report to a military intelligence training school.

Now assigned to the Eighth Armored Division, Salomon's knowledge of the German language gave him the ability to ferret out as much information as possible from the German POWs who were captured as the division advanced to attack each new enemy position.

By 1945, the U.S. troops in Europe knew that the war was drawing to an end. But Hitler's SS, and remnants of his Afrika Korps, still slowed the Thundering Herd's advance, inflicting heavy casualties and destroying many of the division's tanks. Information about German troop concentrations, their mining of bridges and roads, and the locations of food and fuel supplies became essential. But securing this information proved to be no easy task, and Salomon found it necessary to use a variety of means to get the results he needed. Some of his techniques placed him safely within the parameters of the Geneva

*This name has been changed to protect the involved parties.

Convention; others did not. In his interrogation of German captives, Salomon employed threats, offered sympathy, and made promises of early release. He gave men food, water, and cigarettes. But when those tactics failed, as they often did, the interrogator resorted to physical violence. At stake, Salomon believed, was information that could save the lives of his men. He knew firsthand what the Germans were capable of. In addition to his own beatings suffered at the hands of the SS, Salomon had just learned that his mother had been killed in one of Hitler's death camps. For good reason, Salomon made sure his interrogations were witness-free.

One night, in mid-April 1945, as the division stopped on a road, Salomon approached two newly captured German prisoners behind a Sherman tank. Suddenly, Chaplain Vaughn MacArthur appeared at his side. Salomon knew Vaughn to be a kind man and a great help when needed, but at the moment he certainly didn't need the help of a chaplain. Time was short, and Salomon needed enough latitude to interrogate the German POWs as he saw fit.

Vaughn didn't speak German, but he stood close to the trio and listened intently. When Salomon finished his interrogation—albeit a less effective one than he would have preferred—Vaughn interjected.

"Ask them if they know of a prisoner-of-war camp in the area."

Salomon translated Vaughn's request, but both men shook their heads. "*Nein.*"

Vaughn nodded, thanked the interrogator, and left.

A few days later, the Eighth Armored Division had advanced roughly ten miles deeper into Germany. After capturing a German patrol, Salomon commandeered an abandoned farmhouse to use for the interrogations. Somehow, and in a turn of events that would remain a mystery to Salomon for the next fifty years, Chaplain MacArthur again appeared. Salomon did the best he could under the circumstances to interrogate his prisoners under the watchful eyes of the man of God. When he finished, Vaughn asked him to again inquire about possible prisoner-of-war camps in the area. This time, the answer was not *Nein.*

"*Ja,*" one of the Germans said. "*Nicht weit von hier.*" Not far from here. Salomon pulled out his map and asked the men to point out the camp's location.

"At our rate of advance," Salomon said to Vaughn, "we should be there in a few days." Vaughn thanked the interrogator and walked away.

At the time, Salomon didn't know it, but for two months, since he'd first set foot in Germany on March 2, Vaughn had been searching for his son. The lieutenant colonel's rank and reputation had gained favor with the higher-ups, and he enjoyed a certain measure of freedom as the Herd made its way through western Germany.

There were plenty of camps to check. During the thirteen years between 1933 and 1945, the Nazis and other Axis powers operated more than forty-two thousand camps and other areas of incarceration in Europe alone. These included ghettos, concentration camps, forced labor camps, transit camps, killing centers, and prisoner-of-war camps. Many of the forced-labor camps had vast networks of subcamps—some as many as a hundred.

Five hundred of these Nazi camps were designated as POW camps. If Dave were still alive, Vaughn knew he'd be imprisoned in one of those camps. Every time Vaughn learned about a new POW camp, he took his jeep and a translator into the area to question the camp's officials about his son. So far, he'd come up empty.

CHAPTER 24
STARS AND STRIPES

In 1861, shortly after the Confederate States of America first bombarded their neighbors to the north at Fort Sumter, South Carolina, a group of Union soldiers found themselves encamped in Bloomfield, Missouri. They discovered Bloomfield's newspaper office abandoned and began printing their own newspaper that chronicled their wartime activities.

The soldiers called it the *Stars and Stripes*.

The publication lapsed for several decades after the Civil War until, during World War I, it briefly resumed. During the Second World War, the *Stars and Stripes* experienced a second rebirth, with dozens of editions printed and distributed to members of the U.S. military across several theaters of operation.

As U.S. troops fought their way across Germany in spring 1945, two jeeps full of writers from French editions of the *Stars and Stripes* set out in search of printing facilities east of the Rhine River. On April 1, they struck gold in the German city of Pfungstadt, just to the west of the low-lying Odenwald mountain range and some 170 miles from Göttingen, Germany.

The writers worked swiftly. Within four days, they had produced ten thousand copies of their first issue. Within a month, the Germany edition of the four-page paper was being circulated each day among more than a half-million troops in the surrounding area.

The Thundering Herd had come to rest in Göttingen on April 25, 1945. The city would be their headquarters until the first week of June. Some officers settled into local houses while other men billeted elsewhere in the city and surrounding area.

On the morning of May 2, a Wednesday, Vaughn had just finished writing a letter to Dorothy, who was back home in Boston. For U.S. troops scattered across Europe, the day had already proven to be a thrilling one. That morning, the *Stars and Stripes*, in a screaming headline that nearly filled the entire front page, announced the historic news.

Hitler was dead.

In a bunker, above which a parking lot would later be laid, a self-inflicted bullet had ended the Führer's six-year quest to expand the Third Reich and create a unified Europe. After his death, the copyright holder of his bestselling book, *Mein Kampf*, prohibited its printing and distribution. Recent studies estimate that the total number of victims who died under Hitler's regime was as high as twenty million.

As Vaughn turned from his typewriter, he picked up an unread copy of *Stars and Stripes* from earlier that week. Buried deep within the last page was a short article that nearly stopped Vaughn's heart. On April 29, General Patton's Third Army had liberated a POW camp that contained thousands of Allied prisoners in Moosburg. Many of the prisoners were American airmen.

Vaughn grabbed a pair of scissors and cut the announcement from the newspaper. He shoved it into the envelope along with his letter to Dorothy and bolted from the room in search of Lieutenant Colonel Howard Peckham, the Division Chief of Staff. Vaughn needed a plane. He needed a pilot. And he needed to get to Moosburg. He spoke to Colonel Peckham, cleared his schedule, and secured permission from Colonel Henry Holt, commander of the Eighth Armored Division Artillery, to use one of his planes.

By 12:30 P.M., Vaughn and a pilot named Major Cross were well on their way. Clad in a tanker's uniform, an M-1 steel chaplain helmet, and a pair of sunglasses, Vaughn climbed into the observer's seat of the Stinson L-5 Sentinel. Bundled with him in his seat was a large package of cigarettes.

Nicknamed the Flying Jeep, the L-5 was a small, unarmed liaison aircraft that could hold, in tandem seating, one pilot and either a passenger or stretcher. She couldn't carry more than 471 pounds of added weight. If Vaughn were somehow able to find his son, the plane would present two problems. First, it would be an impossibly tight squeeze for the three men, and second, the weight of her passengers would likely impede the plane's ability to take off.

Guided only by a German road map, Vaughn and his pilot lifted off the runway in Göttingen and headed due south above the Thuringian Forest, a forty-mile-long mountain range contoured by steep, sloping peaks.

The flight to Moosburg was just over 250 miles. With a max speed of only 163 miles per hour, the Flying Jeep would take three and a half hours to span the distance. At times, the weather cooperated, but as they passed through the mountains, the tiny aircraft encountered several snowstorms and a pocket of torrential rain.

The skies were mostly free of other aircraft, but they did run across one German plane. The Luftwaffe pilot saw the L-5 and, not realizing the Americans were unarmed, tried frantically to escape. Vaughn watched as the German pilot lost control of his plane and crashed.

Before long, below a sunny stretch of cloudless sky, the men spotted what appeared to be a large prison camp situated two miles shy of the city of Moosburg. Major Cross adjusted his airspeed and prepared to land.

Three days earlier, April 29, 1945, Stalag VII-A, Moosburg

At 10:00 A.M., following failed negotiations for a peaceful transfer of power from the Germans to the Americans, General Patton decided to seize Stalag VII-A by force with the Fourteenth Armored Division of his Third Army. For two and a half hours, Dave and his buddies listened to the war rage beyond the walls. Within thirty minutes of the battle's end, Combat Team A arrived at the camp's gate. The German guards surrendered, the prisoners cheered, and after more than half a year in captivity, Dave realized that he would soon be a free man.

The camp at Moosburg was not the only one to fall in Germany on April 29. Later that same day, some thirty miles to the southwest, the U.S. Army would also liberate the death camp of Dachau.

As Stalag VII-A fell into the hands of the Allies, the Fourteenth Armored Division began the arduous task of processing the tens of thousands of POWs overflowing the camp. While the prisoners waited, they sent a flurry of V-mail messages to their families back home.

Dear Folks,
    The Allies pulled through and we are back in the
Army once more. I have not heard anything from you in

almost 10 months. I imagine that Dad is overseas. It is
pretty rough not hearing anything. We are all well &
have started to eat. The G.I. food sure looks good after 7
or 8 months. This has been quite an experience. We had
quite an exhibition the other day put on by our boys. I
hope that we will be on our way home soon. I am in
good health so don't worry. Will keep you notified as I
progress. It won't be too long. Have had a long time to
plan so I have got some good ones. Have been worried
about you not having heard since I was shot down.

Love,
Dave

Vaughn and Major Cross circled the camp from within their
Flying Jeep, looking for a place to land. Stalag VII-A was situated in a flat valley surrounded by hills, and since there was no
runway in sight, the pilot guided the aircraft down to the field
at the southeast corner of the POW enclosure. Thousands of
faces watched with curiosity as the plane came to a rest.

Vaughn stepped onto the ground. U.S. officers and GIs of
Patton's Third Army swarmed the plane. Vaughn introduced
himself and explained the reason for his visit. A man offered to
show him to the camp command post.

Vaughn followed the man along the back side of the enclosure. His eyes lingered here and there, looking through the
fence and beyond the barbed wire. He scanned the sea of skele-
tons.

Within seconds, the chaplain stopped cold in his tracks. His
eyes locked on a mop of bright red hair, and Vaughn choked out
the only three words his throat could manage:

"That's my boy!"

PART III

No more let Life divide what Death can join together.
—Percy Bysshe Shelley,
Adonaïs

CHAPTER 25
MEATLOAF AND APPLE PIE

"Dave! . . . Dave! . . . *Dave!*"

Dave elbowed through the masses of men and swiveled his head, searching for the voice that seemed to be shouting his name. It was coming from the eastern fence to his left. He pushed his way from the latrine, through the clogs of prisoners, following the sound to its source.

"Dave! . . . *Dave!*"

Through the barbed-wire fence, Dave saw a man dressed in a tanker's uniform. He was wearing sunglasses and a steel helmet. Then Dave saw the cross. The silver crucifix was soldered to the helmet and flashed like lightning in the blinding afternoon sun. At that moment, Dave knew the man was a chaplain. And he knew the chaplain was his father.

In disbelief, Dave stumbled toward the fence, accidentally walking straight into the barbed warning wire. Keeping his eyes locked on his father, he quickly disentangled himself and made a mad dash for an opening in the fence.

Vaughn, too, was now on the move. He raced for the fence,

squeezed through the opening, and with seven months of tears welling up in his eyes, Vaughn pulled Dave into a rib-crushing embrace and refused to let go. Dave buried his head in his father's shoulder and threw his thin arms around his neck.

Vaughn had crossed an ocean and a continent. Dave had witnessed the horrors of holocaust. But at long last, the distance between father and son was closed, and the two men were finally together. The chaplain had found his boy.

Dave took his father back to his tent and introduced him to his buddies. Vaughn passed out the cigarettes he'd retrieved from the plane then went to the CO's office to work on getting Dave released. In less than an hour, Vaughn had secured his son's freedom.

By early that evening, Dave was packed and ready to go. But the sky had other plans. Neither Vaughn nor Major Cross liked the looks of the rain, which advanced in waves toward the camp, and postponed the return flight until the following day.

That night, Vaughn tried to sleep on the ground inside his son's tent with nearly three hundred POWs, but the chaplain was too excited to even close his eyes. Knowing his son was safe was enough. "It seemed so good to be able to look right over and know he was there and sleeping like a log," Vaughn wrote to his wife two days later.

Early reports of the capacity of Stalag VII-A at the time of Dave and Vaughn's reunion put the number of POWs at roughly twenty-seven thousand. In actuality, because of the massive influx of prisoners from the winter marches, by the end of April 1945 the compounds housed far more men than anyone had imagined. On the afternoon of May 2, as Dave made his way from the latrine back toward his tent, he was one of as many as 120,000 POWs in the camp.

Dave drifted to sleep with his dad by his side and a grin on his face: a trip to the latrine in Triolo had gotten him into this mess, and wouldn't you know it, a trip to the latrine had gotten him back out again.

At six the next morning, Chaplain MacArthur, Major Cross, and the newly freed Lieutenant MacArthur were driven by jeep to a field approximately a mile away where Major Cross had stowed their plane. As the men climbed aboard the L-5 Flying Jeep, all three had their doubts.

Major Cross slipped into his pilot's seat. The cockpit was too small for both Dave and Vaughn to sit, so they stretched out like litter patients behind the pilot, lying on their sides so both men would fit. The aircraft's engine roared to life, but for all its noise and even at full throttle, the Flying Jeep budged not even a little in the soggy field. Even though Dave had dwindled down to a mere 135 pounds, the bird was still too heavy.

Instead of a Flying Jeep, the men decided that a regular one would have to do. Dave and Vaughn climbed out of the L-5 and, with their pilot and driver, scoured their German road map. The town of Landshut had an airfield only ten miles to the east. Without the added weight of his two passengers, Major Cross had a plan. He would try to take off from the waterlogged field, fly the plane to Landshut, and rendezvous with Vaughn and Dave in hope of attempting another takeoff, this time on firmer ground.

His plan was a success. The much lighter L-5 inched forward in the grass, gained speed, and left the ground. The MacArthur men watched it rise then climbed back into the jeep and made the short drive to Landshut. By 10:45 A.M., the trio was airborne. For nearly four hours, father and son lay side by side, face-to-face, in the back of their Flying Jeep.

*   *   *

The plane touched down in Göttingen on the afternoon of May 3. Between feeding Dave, delousing him, and securing him a haircut, Vaughn scribbled a hurried V-mail message to Dorothy. He knew his wife was alone back in Boston, so he chose not to worry her by mentioning that Dave was little more than skin and bones.

Dearest
    I found Dave at Moosburg yesterday. He is fine—no loss of weight. Will be with me for a few days.
    Write more later

                        Love
                        Vaughn

Hoping one way or another to get the good news to his wife, Vaughn sent the same message to Dorothy via airmail and briefly considered sending word through an encoded Expeditionary Force Message cable as well. Vaughn went to the office but found it overrun with men sending Mother's Day messages to their wives and moms back home. Besides, he suspected the cable would likely cause Dorothy more confusion than clarity, so he left the office. As he stepped from the building, Vaughn closed his eyes and asked God to help his wife rest easy that night, knowing that her son was at last safe with his father.

Vaughn retrieved Dave from his delousing. Before he finished getting him cleaned up, though, there was someone he needed to meet.

In the town of Göttingen, the interrogator, Technical Sergeant Salomon Braus, had made himself comfortable in a liberated home. Here in the Harz Mountains, the Eighth Armored Division had been resting and giving their Russian Allies time to cross the Elbe River and take Berlin. The house he'd found had everything he could possibly need—hot water, a warm bed, and plenty of food. Most importantly, though, it offered him space to conduct his interrogations of captured Germans.

Early that evening, just as he stepped out of the shower, Salomon heard a knock at the door. To his surprise, he saw Chaplain MacArthur standing before him. It had been at least a week since he'd last seen him, but this time Vaughn wasn't alone. Beside him stood a young, unshaven U.S. pilot.

"Do you know who this is?" Chaplain MacArthur asked. Before the interrogator could say a word, Vaughn answered his own question. "This is my son, my boy who was shot down in a raid, and who has been listed as missing."

Salomon couldn't speak.

"I've hoped and prayed I might find him alive in one of the POW camps," Vaughn said. "With your help as I searched, my prayer has been answered."

Salomon gave the two men a last salute as they climbed into the jeep and drove away.

For the first time in seven months, Dave had a chance to write his mother a proper letter.

Dear Mother,

Yesterday afternoon I was walking from the latrine to my tent in Stamlager 7A when I happened to look out of the barbed wire and saw a sight that I will not soon forget. I was speechless to say the least. We could not say a word for it was quite a meeting for father and son in a prison camp in the heart of Germany.

The tears came to dad's eyes as well as mine, and it did not take us long to get together. It was the first word or news that I had received from my family since I was shot down, and it was worth all of the waiting.

I think that I can truthfully say that these are the happiest days of my life, and they will be that way the day I meet you. I want you to take care of yourself and keep your chin up until I get to you, for I will be on my way with all possible speed. I have been looking forward to this day for a long, long time.

Tonight I am feeling really fine. At last I am clean again and free of everything that would remind me of the past. We are in a warm room in a fairly good house, and it is really the first time that I have been warm in a long, long time. I am really thriving on the best food in the world, good old G.I. grub.

It is all like a dream. For seven long months I wonder what has happened to my family and how they took the news and every detail that goes with it, and you can imagine how I felt when I saw it walk up to that barbed wire yesterday.

Put the date May 2, 1945 down, for it was a great day for the MacArthur family.

Before Dave went to sleep, his father had one more surprise in store. For Dave's entire life, he'd known his dad to be a teetotaler. He was a religious man, a solid man, and he liked to practice what he preached. But that night, all the rules were thrown to the wind, and Vaughn handed his son a glass of whiskey.

The next morning, Vaughn sat down at his typewriter and composed a letter to Dorothy about finding their son. "Gee, I am so happy that I cry rather easily when I think of it," he said. "You can expect it to be between a month and two months before he will be home with you, but I can vouch that he is in great shape. He looks better than that morning when we put him on the train at New Orleans."

* * *

Vaughn was determined to make each day with Dave count.
They went to the movies and played baseball together, but
mainly they ate. Dave needed to compensate for seven months
of lost calories, and feeding Dave became Vaughn's new priority. Soon, he had another update for Dorothy.

This noon we were invited to have dinner with the
general, so we left a good meat dinner and went to his
mess to have spaghetti and meatballs and pineapple. So
when we broke away at 1.15, I took him down to our
mess and there we had meatloaf and string beans and
apple pie and that fixed both of us up in great shape. If
the general ever knew we had chiseled two meals that
day, he would blow his top.

We are going over to the Quartermaster tomorrow to
get some stuff given to Dave, and then I am going to try
to get him a watch and a pen and such stuff like that so
that he will be fairly well off when he leaves here. I am
feeding him fruit juices and plenty of stuff that he likes.
He wanted some peanut butter, so I must get him some
of that somewhere. I am also giving him the camera and
the film to take back.

I hope that my letter gets to you right away, because
that is your Mother's Day gift. I got your son for you,
and I know there is nothing more that you would want
than that. I hope by next Mother's Day the old man will
be back home to harass and trouble you, for he kind of
misses the home ties.

On the back of his father's letter, Dave added another note
to his mom.

Dear Mother,

I am really getting back into shape now and it sure is doing me a lot of good. I have been catching up on my reading that I lost out on and it sure looks good. We have gone to the movies twice now and you know how I like that. Dad and I had a chance to have a long talk last night all alone and it sure brought back the old memories.

I am just going to make this short for I really don't know of much to say. I am having to learn a lot of things over again. My spelling and writing have gone to pot. I guess that I am getting a little better on it now though.

I want to stick around here as long as I can with dad and I want to get home to you as fast as I can so it is hard to know what to do. I think though that I will be well taken care of and when the time comes I will be on the way. Right now if I left I would just spend the time in some camp in England awaiting shipment so I guess that we will wait until traffic thins out.

Dad was telling me of the situation up where you are now and I got all of the latest poop on the rest of the family. Well I guess that I will finish this and get it off.

Love Dave

During the two blissful weeks Dave spent with his father, they shared stories of their wartime experiences and dreamed of the bright future awaiting the MacArthur family. Dave became somewhat of a celebrity among the Thundering Herd. Underneath a large photo of a beaming Vaughn handing a bar of soap and other necessities to his smiling son, the *Armored News* commemorated the remarkable story of Dave's liberation with a press release entitled "Follows Hunch, Finds Son." The caption read: "After delivering his son from a prison-of-war camp, Chaplain (Lt. Col.) Vaughn H. MacArthur of the 8th Armored Division presents him with soap and clothes. The chaplain spotted his son, Lt. David W. MacArthur of Brighton, Mass., in a crowd of 30,000 after flying to the enclosure on a hunch that he might be there."

For Vaughn, his chaplain duties were still a priority. During the month of May, all of which was spent in the Harz Mountains, Vaughn conducted eight weekend worship services and four midweek worship services. He recorded intentional conversations with sixty-three different people, not counting the five welfare cases and four disciplinary cases he handled that month. He also made four hospital and aid-station visits, during which he talked with more than twenty soldiers. The war was screeching to a halt, and there wasn't much fight left in the Nazi machine. But the Herd was still in a war zone, and there was plenty to keep Vaughn busy.

As units of the Eighth Armored Division were working overtime to relocate former Russian POWs and displaced persons, Dave made himself useful. Working with the interrogators, Dave was able to help sniff out nearly a dozen German war criminals. Three of them were convicted of atrocities warranting death.

It felt good to be on the other side of the table for a change.

Shortly after midnight on May 7, 1945, less than a week after Dave's liberation, Germany surrendered. Like all wartime updates since the invasion of Normandy nearly one year earlier, V-E Day was announced the next morning in the daily communiqué from SHAEF, the Supreme Headquarters Allied Expeditionary Forces in Europe. Within hours, U.S. troops scattered across the Continent could read the good news for themselves in the various editions of the *Stars and Stripes*.

"UNCONDITIONAL SURRENDER IS ANNOUNCED BY GERMANY."
—Southern France

"IT'S ALL OVER OVER HERE." —Rome

"VICTORY"—Paris

"NAZIS QUIT!" —Germany

In the streets of Göttingen, father and son celebrated the Allied victory. The war in Europe was over.

After spending two weeks with his father, Dave bade farewell to the men of the Thundering Herd. Clouds blanketed the sky as Vaughn and his driver shuttled Dave from Göttingen to the Halle Airdrome, a little more than a hundred miles across Germany. Dave would soon be in the air again, boarding the first of several flights across the Continent. Along the country road, they passed by a crashed German Me 109 resting in a field. The two men hopped out, snapped a photo with the broken bird, then continued on their way.

When they arrived at the Halle Airdrome, Dave slid out of the jeep and hugged his dad good-bye. A ray of sunlight broke through the clouds, spotlighting his father as he drove away into the distance.

CHAPTER 26
HOME

Dear Mother,

Am on my way home. Will be seeing you in a few days or weeks. Am in good shape. Never got a letter while here. Am eating like a king. Have to get this on its way. Hope that everyone are well & happy. Hope they are all in the States. Will be seeing you soon.

Love
Dave

Dave scribbled these words on May 21, just before starting his journey home. He would spend the next several weeks working his way across Europe through the POW repatriation process before boarding a ship for the States. Once he disembarked in America, he would be sent to rest and recuperate, spending a full thirty days at home in Boston with his mother and younger brother.

Halle, Germany, had been under Allied control since April 17 when U.S. soldiers overtook the city. Dave was delighted to find that the Halle Airdrome's hangar had become a storehouse for planes of all sorts. He even had time to fly a few.

He took up a Fieseler Fi 156 Storch. The German liaison aircraft known as the *Stork* was similar to the L-5 he and his dad had flown from Moosburg back to Göttingen. But this time, Dave was in the pilot's seat. For the better part of a year, the flyboy had been grounded, but now he was back up in the air where he belonged.

Soon, it was time for Dave to depart. With several stops in between, he was flown from Germany to Camp Lucky Strike, one of the U.S. Army's many "cigarette camps" set up near the port cities of northern France, not far from the beaches of Normandy. There, Dave was processed and then waited with other freed POWs in the temporary tent city for a liberty ship to take him through the English Channel and back across the Atlantic.

Dear Mother,

We sail in the next day or two. I don't know where we land yet. It will be either Boston or New York. It should be between the 10th and 12th of June. The ship is the S.S. Lawson. I have been so busy that I have not been able to write. I will let you know just what is up when I get to my first phone. Don't worry, I will be home soon.

Love,
Dave

June 9, 1945, Pilsen, Czechoslovakia

Three days after the Eighth Armored Division arrived in Czechoslovakia to break up firefights between the Czechs and the Russians, a jeep hurried down the shell-pocked roads of the city of Pilsen. The vehicle lost control, careened off the road, and rolled over. Its passenger, who had just completed a series of hospital visits in the area, was trapped underneath. It took several men to lift the wrecked jeep off the man's crushed chest. Once freed, he was evacuated to the hospital he'd left only minutes before. This time, he was a patient.

At the hospital, doctors performed a thoracentesis, inserting a needle into the space between his chest wall and lungs to remove excess fluid. But the damage was too extensive. Within hours of arriving at the hospital, the man succumbed to his injuries and died.

A few days later, the SS *John Lawson* arrived in Atlantic City, New Jersey. For the first time in nearly eleven months, twenty-year-old Dave was back in America. Having completed his seafaring voyage and his service in the Second World War, he set foot on terra firma and made his way to Boston. When his mother, Dorothy, saw him, she threw her arms around Dave's lanky frame and wept. Her son was finally home.

Days later, on a warm New England afternoon, Dave arrived back at his mother's house to find a notice affixed to the front door. It had been left by a bicycle messenger whose knocks had gone unanswered.

Dave opened the envelope and read the notice. A telegram was waiting for Dorothy at the Western Union office. He rushed to his car and drove to the office. After presenting his identification, Dave was handed the telegram. The words on the thin slip of paper buckled his knees.

THE SECRETARY OF WAR DESIRES ME TO EXPRESS
HIS DEEPEST REGRET . . .

His father was dead.

*   *   *

The day before Vaughn's fatal jeep accident, he had received word that Dave was officially sailing home from Camp Lucky Strike. Vaughn had lived up to his promise. He had freed his son, and he died knowing his boy would be okay.

On the day that would have marked Vaughn's third anniversary with the Eighth Armored Division, the Thundering Herd honored their beloved chaplain in a memorial service held in a local Pilsen church. Fifteen years had passed since Vaughn's first appointment as a chaplain in the U.S. Army. He was only forty-five years old.

Major General Devine, the Herd's commander, and other officers and men of the Eighth Armored Division attended Vaughn's funeral. In the division's written history, Chaplain MacArthur was eulogized with the following words:

> The entire Division was saddened by the accident in which Chaplain Vaughn MacArthur lost his life. His untimely death was a loss to the Division at a time when spiritual values needed re-emphasis. Known as a front-line chaplain, he had made frequent trips under enemy fire to minister to combat troops and to the sick and wounded. Possessed with an ever-cheerful personality and a calm, confident attitude, Chaplain MacArthur was an inspiration to officer and enlisted men alike.

The Army's Acting Chief of Chaplains, Brigadier General Luther D. Miller, offered his condolences to Vaughn's grieving widow.

> My dear Mrs. MacArthur:
>
> I am so sorry that I was not in to see you personally when you called at the office the other day. May I

express the deepest sympathy of the Chaplains Corps, which shares with you the loss of your husband.

It will be comforting to you and your loved ones to know that the men whom Chaplain MacArthur so faithfully served hold him in the highest esteem and respect because of his work and example as a minister of God. His loyal and sacrificial devotion to his Country and his Church will be remembered long by all who knew him.

It is our prayer that the Heavenly Father will give you strength to bear the heavy burden that is yours, comforting you in your sorrow with His everlasting love. I shall remember you in my own prayers.

Sincerely yours,
Luther D. Miller
Chaplain (Brig. Gen.) USA
Acting Chief of Chaplains

Dorothy would save that letter for the rest of her life.

For his service in the Second World War, Lieutenant Colonel Vaughn Hartley MacArthur was posthumously awarded the Bronze Star. His body was interred in Plot G, Row 16, Grave 22 of the Lorraine American Cemetery in Saint-Avold, France, with ten thousand other brothers in arms at his side.

Like his son, Vaughn, too, was finally home.

CHAPTER 27
BUGOUT

The morning of April 22, 1951, behind enemy lines
Wontong-ni, South Korea

It had been nearly six years since Dave returned home from Europe, and he was once again at war—this time on the other side of the world.

Dave surveyed the army of Chinese soldiers for a spot through which to plow the front of his jeep. Dave had been captured once by the Nazis, and by God, he wasn't going to be captured by the Chinese, too. In those fleeting, eternal seconds, as Dave prepared for the most dangerous ride of his life, his mind jumped back to another time, to another place, to another man who had died in a jeep.

What felt like a bolt of lightning coursed from his head to his heel, and with a violent surge, he clenched the wheel, punched the gas, and launched the convoy into motion. The twenty jeeps burst forward, surging recklessly at the charging wall of Chinese soldiers. Dave slammed the clutch, shifted gear, and then pounded the accelerator. Everything in his body was starving for speed.

Seen from above, the convoy looked like a narrow battleship effortlessly cutting through a wave. But down below, with Dave at the bow, the image was anything but elegant. First came the splattering of bullets, then the horrified expressions on the Chinese faces. At the moment of contact, with every muscle in Dave's body fully tensed, the scream of his engine was suddenly matched by the crack and smack of bodies slamming against his front fender.

The jeeps plowed headlong into the mob, penetrating the wave and creating a loud percussion of thuds, thumps, scrapes, and screams. Bodies were sucked below and crushed by the weight of the wheels. Limbs were torn from torsos. Burp guns, helmets, blades, and bugles burst into the air only to be gobbled up by the jeeps bringing up the rear. Dave kept his chin down and his head low. He tucked his arms tightly to his sides, avoiding the clawing hands reaching through the doorless vehicle.

The carnage continued, but Dave didn't ease up. He'd lost a lot of men during the attack on his village. This ride was for them. It was also for the men now looking to him for salvation. Dave himself had once been saved, and now it was his turn to return the favor—even if his wheels had to paint the road with blood as bright as the Red Army's flag.

The mob lunged at the jeeps, stabbing and pulling men to the ground. Other ROKs fell out and were instantly swarmed by knife-wielding enemy troops. The convoy was moving more slowly now. Ruthless rounds of ammunition peppered the jeeps.

One of the bullets had Dave's name on it. The lucky piece of lead came hot through the windshield and grazed the underside of his chin, missing his throat by millimeters and leaving a deep furrow in his skin. Dave gunned the gas harder, desperate to break free and put more distance between his jeep and the overtaken village of Wontong-ni. In the rear of the convoy, the Chinese had already capsized several jeeps. Another barrage of bullets killed a Korean in the jeep immediately behind Dave, tossing the man's body from the vehicle.

Just then, a familiar sound joined the cacophony. Dave craned his neck to see the wings of a T-6 Mosquito rise up from behind the convoy. Never had a World War II trainer looked so majestic. The pilot began showering the Chinese with shells. Dave saw the Mosquito swoop overhead, guns blazing, engine roaring. The pilot banked for another strike and continued thinning out the troops on the road to allow the remaining jeeps to break free. When the road was clear, the Mosquito fell back, loitering above the convoy in lazy circles. Dave eased up on the accelerator. Only ten of the original jeeps were still behind him.

Suddenly, Dave's passenger, Captain Park, slumped against him. A bullet had found a place between the interpreter's eyes. Under the weight of the man's body, Dave's jeep swerved to the left. He nudged the man with his arm, but the body slumped over a second time. In an effort to regain control of his vehicle, Dave elbowed the man harder. To his horror, the lifeless interpreter tumbled from the jeep and onto the road.

Dave looked back to see the vehicle behind him swerve, strike the corpse, flip, and catapult its passengers into the air. Two additional jeeps were demolished when they collided with the overturned vehicle. Dave slammed his brakes and ran to the wounded men. Helping them to their feet, he loaded three of the men into his jeep and continued driving south on Route 24.

They passed the destroyed village of Soho-ri, which was

still swarming with enemy troops. The Chinese saw the six remaining jeeps approaching, and they unleashed a salvo of small-
arms fire at the convoy. Dave reached for his portable radio.
He called up the Mosquito pilot to see if the road was clear.

"You'll have to break up the roadblocks," the pilot's voice
crackled back. "There are at least three immediately south of
your position." The Mosquito destroyed the first roadblock,
but with more urgent targets to pursue, he peeled away. "I wish
I could pin a medal to that pilot," Dave later said, "but I don't
even know who that guy was."

The abandoned village of Ip'yong-ni was now fast approaching. Dave drove across the Soyang River bridge and
parked the convoy behind a cluster of buildings. The Chinese
were firing from the hills, so the men scrounged up what little
ammunition the South Koreans had left and hunkered down
for the firefight.

Dave and an Australian American "digger" spotted two
M20 recoilless rifles on the ground, made a beeline for the
abandoned weapons, and began firing phosphorus shells into
the hills at the Chinese positions. Eventually, the ammunition
ran low and Dave rushed back to his jeep radio, hoping to call
down an air strike on the ridge. But when he turned the corner
of one of the buildings, he stopped dead in his tracks. Staring
back at him was an entire patrol of Chinese soldiers. Neither
Dave nor the fifteen armed men standing a few feet in front of
him were expecting the sudden face-off. "I can remember to
this day," Dave later said, "the leader of the patrol had a gun in
a holster, and he had it wrapped in cloth, and he was madly
clawing for that gun."

Without thinking, Dave slipped his hand to his hip, un-
snapped his leather holster, and drew out his .45-caliber pistol.
The hammer was already cocked. A bullet was waiting in the

chamber. Dave had a fraction of a second to make that bullet count. The patrol leader raised his weapon, but Dave outdrew him and "let him have the whole clip."

The Chinese leader sprang backwards, his chest riddled with holes. The other soldiers stood paralyzed long enough for Dave to toss them one of the grenades he kept fastened to his belt and bolt around the side of the building. The frag exploded, and so did the patrol.

Dave grabbed his portable radio and positioned himself high enough in the hills to gain a visual on the Chinese units moving across the river. He reached an F-9 Cougar, which dropped napalm along the ridge. After summoning a Corsair to destroy the remaining two roadblocks, Dave scrambled down to his convoy, regrouped, and ripped off in the jeeps.

They followed Route 24 as it curved around the natural bend of the river where the village of Ch'op'yong-ni sat vulnerably beneath a four-hundred-foot cliff. Inje was just around the corner, but large pieces of artillery and abandoned vehicles were clogging the road.

"Across the fields!" Dave yelled.

But as he swerved into a rice paddy, his wheel struck the bank of a ditch and catapulted him, along with his portable radio and his M-1 Carbine, into the water. Two additional jeeps careened into the sludge and were abandoned.

Dave muscled the vehicle back onto the road, and soon the convoy was racing through another sheet of enemy gunfire and mortars. When they passed by Hoch'on, where the road forked, another jeep accidentally veered left onto Route 24E, directly into the enemy-occupied village of Hanyang-dong. None of its passengers survived.

Only three jeeps remained. The convoy followed the road as it took a hard right turn where large groups of retreating ROKs had gathered in the open field near the outskirts of Inje. There were about three thousand men, by Dave's count. Most of them belonged to the Sixth ROK Division that had been retreating south. All the while, the Chinese troops were closing in from the north. They had scrambled up the six-hundred-foot ridges and were positioning their machine guns.

Dave braked to a stop and instructed the other men to continue driving south without him. In the field near Inje, one of the ROK commanders futilely tried to line up his men for a head count, but the Sixth ROK stragglers milled aimlessly. The area was exposed and unprotected.

Dave reasoned there were enough men to form a military unit, so he decided to take charge of the situation. He corralled the ROKs and dispatched forward scouts and sentries to guard the perimeter.

After exchanging a few words with the ROK infantry officers, Dave changed out of his blood-soaked clothes and donned fresh ones. He removed the camera from around his neck and saw, to his surprise, a gaping hole punched through the strap. A bullet had cut clean through it, missing his heart by inches.

Suddenly, hell broke loose. The ROK scouts had abandoned their posts and Chinese troops were raking the field with machine-gun fire. Two hundred Koreans were mowed down in the first blast. Dave darted for his jeep radio. As he ran, he scanned the sky in hopes of contacting a passing warbird. But he was too late. A stream of bullets had ignited his jeep into a blaze of flames.

By this point, the ROKs had reached a full state of panic. They ran for the river. The Chinese reloaded. Dave yelled at them to regroup and defend themselves, but the Koreans were already in the water, swimming frantically for the opposite bank. The machine guns let loose and four hundred more men were slain. The river ran red.

Dave saw that one of the Koreans had dropped a machine gun and ammo in the mud. He picked up the weapon, collected a few magazines, and made his way for the hills. He scrambled up the slope and mounted the machine gun on a rock. With the magazine loaded and gun cocked, Dave dropped to his belly and swiveled the muzzle to face the wave of Chinese troops crashing over the ridge. His camera, now hanging from a pierced strap around his neck, rested gently in the dirt.

Something else also hung from his neck. Seven years earlier, when Dave first arrived at Stalag Luft III, the Nazis had issued him a dog tag to wear. Stamped into the rectangular metal were the words "Oflag Stalag 3" and his POW number, 8754. Dave's hands tightened around his weapon as he remembered the German machine guns perched high in the goon boxes.

Through his gunsight, Dave saw dozens of Chinese soldiers moving in his direction. To inflict the maximum amount of damage, Dave needed them to be closer. He lay completely still and took a few deep breaths to slow his heartbeat. The Chinese grew larger in his field of view. Still he waited.

Just then, at the last possible moment, as the Chinese were almost upon him, Dave braced for the recoil and squeezed the trigger. Nothing happened. Dave's heart rate skyrocketed as he checked the magazine and furiously wiped the mud off the muzzle. The Chinese were nearly close enough to see the panic in his eyes, and one of the soldiers spotted him.

Dave squeezed the trigger again. As the dirt around him kicked up with enemy fire, Dave felt the true horror of the situation. The machine gun was jammed. Dave knew it, and the Chinese knew it.

Dave whipped out his .45 and fired a few slugs, but the enemy had overrun his position. He had to make a quick decision: continue fighting and die, or try to save a few more men. Dave glanced over his shoulder and saw hundreds of Koreans floating facedown in the river below. He didn't know it at the time, but among the three thousand ROKs in the area, only a thousand of them were still alive.

Dave stood to his feet, drew out his knife, and buried it within the rib cage of the nearest Chinese body. He pulled out his bloody blade and sprinted down the hill. Bullets whizzed by, but Dave kept running and didn't stop until he reached the waterline. He jumped into the river and swam for his life.

When Dave reached the bank, a group of ROKs helped him out of the water and turned their fire toward the Chinese descending the slopes. He hurried to assemble the ROKs into smaller units and ordered them to keep a low profile. He hand selected ten ROKs to be in his unit and fled south toward friendly lines.

For the next forty-eight hours, Dave led an exodus away from the enemy. Evasion, not aggression, became the lifesaving strategy. Dave dodged Chinese outposts. He camouflaged himself with mud and painted his body black with ashes. He kept his men walking day and night, avoiding the main roads and sticking only to the footpaths that meandered deep into the Korean mountains.

During the day, the temperature stayed in the mid-fifties, but at night it hovered barely above freezing. The first evening was dry, but on the second, light rain soaked the weary ROKs.

They drank from streams and occasionally received water from hospitable villagers. Once, after Dave gulped mouthfuls of water from a stagnant puddle, he looked up to see a dead body floating near his lips.

It was important for Dave to keep his unit small and stealthy, but stray ROKs kept joining his company as they made their way through the mountains. Before long, Dave was leading a group of two hundred men, many of whom were wounded. Occasionally, when an ROK was overtaken by exhaustion and lagged behind, the sound of gunshots echoed from the hills. The Chinese were never far behind.

Dave's men came under intense enemy fire multiple times every day. One morning, they came within one mile of an entire Chinese army. As they drew closer to the allied lines, Dave heard the sound of 155mm howitzers. "That's how we knew which way was south," he later said. "We were following the sounds of the weapons."

In the early morning hours of April 24, Dave stumbled upon a farmhouse. Using his watch, fountain pen, and all the money in his pocket as currency, he bribed the civilians to guide his men on the fastest route to the allied lines. The plan worked, but the pace was unforgiving. In ten hours, Dave and his men crossed more than thirty miles of Communist-occupied territory.

By 1:00 P.M., Dave came to the bank of a river. He peered across the water to a glorious sight that had been days in the making. On the opposite shore were U.S. Army trucks, driving up and down the road. Helping the wounded ones, Dave ushered his men across the river and flagged down the trucks. Within a few minutes, he loaded all 127 men into the vehicles, destined for the safety of the local U.N. headquarters detachment.

But Dave didn't stop to rest. There were still others trapped behind enemy lines, so he made his way back to a nearby command post where more remnants of the Seventh ROK Division had begun to regroup.

"I have to go south and get more equipment," Dave told a KMAG colonel in charge.

At the time, the Chinese had already surrounded the command post and would soon tighten the noose. In the state of confusion, the colonel agreed. Dave jumped into the jeep and followed the road farther to the south. On the previous day, as the Chinese encircled the area, they had constructed three roadblocks. Dave burst through each of them. He soon arrived at a supply depot and began hunting down a radio jeep and a squadron willing to follow him back into battle.

As Dave was gearing up, a T-6 Mosquito landed on the runway beside the depot. The Mosquito belonged to one of Dave's greatest heroes, a living legend who had flown over the beaches of Normandy on D-Day in a P-38 Lightning. His name was General Earle E. Partridge, Lieutenant General of the Fifth Air Force. In three years' time, he would be commander of the entire Far East Air Force.

General Partridge took an interest in the young pilot and struck up a conversation. As Dave shared his plan to go back north into combat, though, one of the nearby Air Force officers overheard.

"Nobody's going back north," the officer interjected. He took one look at Dave and ushered him to the infirmary.

"What about all our people up there?" Dave protested.

"They'll just have to fight their way out."

Over the next week, Dave was treated for a number of injuries: a fractured jaw, a bullet wound beneath the chin, dysentery, exhaustion, and severe dehydration. But each day of his stay, Dave still demanded to go back to the front lines and continue to serve his country as a Forward Air Controller.

In May 1951, the allies regained control over the lost Korean territory. The Chinese had failed to capture Seoul and were pushed back above the 38th Parallel, which would crystallize into the boundary separating South Korea from the North. The Communists' dream of reunifying Korea was dashed.

Approximately two years later, and without detonating a single atomic bomb, the Korean War came to an end. A Third

World War was avoided, but it came at the price of thirty-seven thousand American lives.

The Chinese Fifth Phase Offensive was the largest battle of the Korean War. From April 22 to 29, the Chinese suffered somewhere between thirty-five thousand and sixty thousand casualties. Dave MacArthur was credited with killing over a thousand of them. He had spent seventy-two hours behind enemy lines and saved the lives of 127 men.

Later in his life, Dave was asked if he felt like a hero. "No," he replied. "I was just saving my ass like everybody else." But the United States begged to differ. On December 19, 1951, President Harry S. Truman awarded to First Lieutenant David W. MacArthur the Distinguished Service Cross. Dave was one of only forty-two Air Force recipients of the nation's second highest award for military valor.

The President of the United States of America, under the provisions of the Act of Congress approved July 9, 1918, takes pleasure in presenting the Distinguished Service Cross (Air Force) to First Lieutenant David Warren MacArthur, United States Air Force, for extraordinary heroism in connection with military operations against an armed enemy of the United Nations while serving as a Forward Air Controller, 5th ROK Regiment (Attached), 7th Republic of Korea Division, in action against enemy forces in the Republic of Korea from 21 to 23 April 1951.

After the Fifth Regiment was overrun and surrounded by Chinese Communist forces, annihilation was imminent. Although morale of the men was badly shaken, Lieutenant MacArthur reorganized the group and despite intense enemy mortar, small arms and artillery fire, continued to direct effective air strikes against enemy posi-

tions for several hours. During this period, as he talked friendly fighters into their targets, he was wounded, his radio jeep was destroyed, and his interpreter and radio bearer killed by his side.

Undaunted, Lieutenant MacArthur rallied the disorganized troops and led them from impending disaster. For two days, traveling a distance of fifty miles, exhausted and without food, Lieutenant MacArthur and his depleted force successfully evaded capture and continued to harass the enemy. Although many of his own men became casualties, Lieutenant MacArthur, through resourcefulness in the face of bitter enemy action, kept a small contingent intact and led them to safety.

Following his week in the Korean hospital, Dave was flown to Tokyo to be debriefed. Since he was classified as an escapee under enemy control, the Air Force intelligence interrogated him, and the Geneva Convention forbade him from returning to combat.

Dave spent the entire month of June at sea. With recent first-hand knowledge of enemy engagement, he was assigned the task of training pilots on escape and evasion tactics aboard the USS *Princeton*—one of the same aircraft carriers that had provided close air support for him during his April bugout.

Most days, the *Princeton* launched its Corsairs, Panthers, and Skyraiders on missions over the Korean Peninsula. On June 12, though, severe thunderstorms swept over the Sea of Japan. Air operations were restricted until the unfavorable weather relented.

The night fell, and so did the lightning. And somewhere in the distance, the sky bent down to kiss the stormy sea.

*   *   *

On July 3, the USS *Princeton* docked at the U.S. Navy base at Yokosuka, Japan. Dave and the pilots disembarked just in time to celebrate the Fourth of July. Dave's whole life was summed up in his nation's anthem. He'd seen the rocket's red glare, the bombs bursting in air. And at the port in Japan, not far from where Little Boy and Fat Man had split their war-ending atoms some six years earlier, the flag was still there—waving for Dave, for Vaughn, and for all who belong to the land of the free and the home of the brave.

AFTERWORD
ACE FOR THE ENEMY

On May 1, 1953, a year and a half after Dave received the Distinguished Service Cross for his heroism in Korea, President Eisenhower awarded him the Purple Heart for wounds received in action during his fateful strafing mission over Greece in World War II. It would be one of many awards he would receive over the course of his remarkable military career.

Dave had been the last member of his family to see his father alive. The news of his death, Dave said, "pretty well finished the job that the Germans had almost succeeded in doing." Two months after he returned from Europe, Dave transitioned from active duty to the U.S. Army's Officers' Reserve Corps. He was now the head of his household and had to shoulder the weight of providing for his family. "Besides that I had to get back into college," he said. "It had been the ambition of my father to see that all of us got through college, and he worked and died for that."

Dave picked up his studies at Duke University, where a fellow student published a riveting article about his experiences overseas. His stay in Durham, North Carolina, was short-lived. Following a tour of "a full gamut of colleges," Dave finally graduated in 1950 from Bates College—his father's alma mater.

A year after he returned from World War II, Dave had a chance to think deeply about his experiences. Many of them were recorded in a rust-colored notebook he labeled *Miscellaneous*.

His memories, which returned in spurts, frightened him. "I figured that I had had my run of luck and that I was living on borrowed time," he wrote. "I also tried to figure out how anyone could be as fortunate as I had been during my imprisonment."

"Was I scared?" he was asked, decades later. "Hell yes, I was scared. Most of the time. . . . The training I had taught me well, and I was foolish enough to take it all to heart. I also had luck, and God."

"I guess what got me through it," he said, "was that I was young, I was loyal, and extremely patriotic. To this day, nothing makes me cry except when I hear 'The Star-Spangled Banner.'"

Dave's experiences in the Second World War were extraordinary, but perhaps the most courageous moments of his life—the moments that refined his character and defined his legacy—occurred in Korea. What happened during the seventy-two harrowing hours of April 21–24, 1951, became known as one of the most heroic episodes in U.S. military history.

Three years after returning from the Korean War, Dave married a smiling brunette named Nancy and welcomed the first of two beautiful daughters—a baby named Pamela who entered the world on Veterans Day. Pamela's younger sister, Kristin, completed the family two years later.

That same year, in the fall of 1958, Dave was assigned to the 100th Bomb Wing at Pease Air Force Base in New Hampshire flying the Boeing B-47 Stratojet. The six-engined, high-altitude, subsonic bomber had been designed for a single purpose: to reach and strike the Soviet Union with nuclear warheads.

But the bomber didn't suit Dave. "I didn't like carrying nuke weapons around," he later said. "If I was going to fight, it was going to be man to man, not dropping bombs on innocent people."

In the fall of 1966, the United States had been entrenched in the Vietnam War for roughly two years. Unbeknownst to his wife, and with his daughters on the cusp of adolescence, Dave quietly volunteered for active duty.

He flew a dozen combat missions over the jungles of Vietnam in his F-105 Thunderchief. But when Lieutenant Commander John McCain was shot down and imprisoned in the Hanoi Hilton in October 1967, the U.S. could no longer risk providing the Viet Cong with "elite" POWs. With his famous last name and former POW status, Dave was no longer allowed to fly combat. He spent the rest of the next year on the ground, serving as the chief of maintenance for his unit in Takhli, Thailand, and received an Air Force Commendation Medal (First Oak Leaf Cluster) for his meritorious service.

Dave rounded out his career commanding a maintenance squadron at Hickam Air Force Base in Hawaii before finishing his service at Tinker Air Force Base in Oklahoma, a mere 130 miles from his old training grounds in Muskogee.

On February 1, 1971, at the age of forty-six, Dave, "having served faithfully and honorably," retired from the United States Air Force. Just like his father some twenty-five years before, he completed his duty and service as a lieutenant colonel. In a military career spanning twenty-eight years, Dave had flown thirty different types of planes, from fighters to bombers and everything in between.

Dave transitioned well to civilian life, leaving the military one day and walking right into his civilian job the next. He spent his second career traveling the globe selling a new generation of flight simulators—upgraded versions of the Link he'd trained in so many years before. These new models simulated Black Hawk helicopters, FA-18 Hornets, and even the entire engine room of a naval destroyer.

By the time of his retirement, Dave had flown for the Army and the Air Force, and had lectured on escape and evasion aboard aircraft carriers for the Navy. He had fought for his country in three wars—and for the enemy in one, if counting his brief stint as a Japanese fighter pilot attacking Pearl Harbor in the 1970 Academy Award-winning *Tora! Tora! Tora!* During the two months of filming, Twentieth Century-Fox used fifty-one planes. Dave was assigned to a T-6 Mosquito modified to resemble a Japanese Zero. "The drunks off the street won't know the difference," a production director said.

Dave's final scene called for his death, a dramatic moment in which the pilot slumped over in the cockpit after being shot. After ten frustrating takes, one thing became abundantly clear:

Dave MacArthur just didn't know how to die.

Nancy MacArthur, the mother of Dave's two girls, passed away in June of 1986. But like so many other second chances life had given him, Dave got a second chance at love.

At 11:00 A.M, on Veterans Day of 1988—exactly forty-four years after Chaplain Vaughn MacArthur commemorated Armistice Day in the waters of the Atlantic—Dave found himself standing at the altar in the United Nations Chapel in New York City. Beside him, in pearls and a beautiful white suit, was his new bride-to-be, Sharon Kinne, the woman with whom he would spend the rest of his days. Witnessing the marriage were

Sharon's two teenaged children, fifteen-year-old Dave and thirteen-year-old Bethanie.

Even in retirement, Dave never lost his fighter's spirit. During their first year of marriage in Cushing, Maine, Dave came home to find his wife's new garden trampled by the neighbor's cows. Dave rushed into the house, grabbed his .45 pistol, and mounted his white Honda motorcycle. An astonished Sharon looked out the window to see her husband firing shots into the air as he herded the three beasts like a cowboy back into their pen.

"He never gave up the idea that he was going to have a great adventure again," his friend John Shenton recalled. "And he hated getting older." After decades in the air, flying on commercial planes bored the flyboy. "I'm not interested in flying unless I can turn the airplane upside down," Dave said.

Dave would, in fact, have relished a calmer flight in the skies above Greece on October 6, 1944. He wasn't supposed to be in the air that day but was picked to take Al Nerney's place. Forty-five years later, at a reunion of the Forty-ninth Fighter Squadron, the two men met for the first time since Italy.

"Al!" one of the other pilots called, waving to his buddy. "Come on over here. Do you know Dave MacArthur?" Al walked over, took one look at Dave, and struggled to speak. A devout Catholic, he had spent four and a half decades racked with guilt, praying daily for the man who had died in his place.

Dave hadn't died that day, but throughout his life, he had experienced some of the darkest horrors of humanity in conflicts covering the globe. Many of those effects lasted a lifetime. He ate quickly. He kept his cupboard packed with cans of hash and deviled ham, and jars of pickles. If administered the sedative Versed during a medical procedure, Dave immediately flashed back to war, flailing his arms and yelling at the top of his lungs.

He also suffered from the lingering physical effects of war. His lungs remained at only 40 percent capacity, damaged first by Nazi gases and then again by Agent Orange. He'd had thirty separate surgeries for carcinomas on his face—skin that spent decades unprotected against the rays that pierced his cockpits.

Dave spent the final days of his life in hospice at his home in Venice, Florida. "Boy, those are pretty mountains," he said to his wife, Sharon, as though the boring beige wall in front of him had transformed into a window overlooking the Alps. "When do you think I'll see my Maker?" he asked. "Do you think I'll see my father?" The family's *second* Lieutenant Colonel MacArthur received his answer when, on January 20, 2017, he lost an eight-month battle against esophageal cancer.

Several years before that final fight, Dave was interviewed about some of the mishaps that occurred during his military career. "I didn't count," the man said, "but I think you lost two planes and four jeeps?"

With his thick waves of hair now a snowy white, Dave raised his finger to the sky and grinned.

"Only one more, and I'll be an ace for the enemy!"

On the morning of Friday, February 17, 2017, an Air Force chaplain performed Dave's funeral. Twenty-one guns saluted the pilot. Two warbirds were scheduled to fly over the Sarasota National Cemetery that day. Due to a recent change in procedure, however, the jets never arrived.

But as Dave's family gathered near the concrete pavilion, something in the sky caught his niece's attention. In place of the absent planes flew two bald eagles, circling the service, paying homage to the fallen flyboy from Boston.

ENDNOTES

Except where noted, all letters to and from Dave MacArthur or his family members are from the personal collection of David W. MacArthur, used with permission from Sharon MacArthur. Additional information is from a variety of interviews conducted near the end of Dave's life, from Dave's 1946 typed account of his WWII experiences, and from various other material from Dave's collection. Wartime action sequences draw from declassified after-action reports and mission air crew reports. Much of the information on Vaughn MacArthur is from Dave's personal collection and from Vaughn's official military chaplain records.

One-Man War

4   Wada Haruki, *The Korean War: An International History* (Lanham, MD: Rowman & Littlefield Publishers, 2014), 170.

5   Korean War News Report, as cited in a 6164th Tactical Control Mosquite News reunion letter for September 20–23, 1990, from Stephen J. Rooney, Sacramento, California. From the personal collection of David W. MacArthur.

8   Robert F. Futrell, *The United States Air Force in Korea, 1950–1953* (originally published by Duell, Sloan and Pierce, 1961; revised 1983, 1991, Washington: U.S. Government Printing Office, reprinted 1996), 363.

11   Billy C. Mossman, *United States Army in the Korean War, Ebb and Flow, November 1950–July 1951* (Washington D.C.: Center of Military History, United States Army, 1998), 390.

12   Anthony Farrar-Hockley, *The British Part in the Korean War: An Honourable Discharge, Vol. 2* (London: The Stationery Office, 1995), 114.

Keep 'Em Flying!

18   "Adolf Hitler: Man of the Year, 1938," *Time* magazine,
     January 2, 1939.

20   "Aviation Cadet Training for the Army Air Forces" brochure,
     available at history.cap.gov/file/1432.

22   Ibid.

23   Robin Olds, *Fighter Pilot: The Memoirs of Legendary Ace
     Robin Olds*, eds. Christina Olds and Ed Rasimus (New York:
     St. Martin's Press, 2010), 1.

The North Pole

37   "Vultee BT-13 Valiant," Western Museum of Flight, accessed
     November 3, 2018, wmof.com/bt-13.html.

43   John H. Finley, ed., *Nelson's Perpetual Loose-leaf
     Encyclopædia: An International Work of Reference,* vol. 5;
     entry updated November 1939 (New York: Thomas Nelson
     and Sons, 1917), 399.
     "Capture Hitler Plan Is Jeered. Million Dollar Reward Offer Is
     Analyzed by U.S. Laws," *The Chester Times* (Chester,
     Pennsylvania, May 2, 1940), 20.

44   John Lukacs, *The Duel: The Eighty-Day Struggle Between
     Churchill and Hitler* (New Haven, CT: Yale University Press,
     2001), 7.
     Volta Torrey, "The War's Most Closely Guarded Secret
     Revealed: How the Norden Bombsight Does Its Job" in
     Charles McLendon, ed., *Popular Science Monthly, Mechanics
     & Handicraft: A Technical Journal of Science and Industry*,
     Vol. 146, No. 6 (New York: Popular Science Publishing, June
     1945), 70.

45   T/Sgt. James Lee Hutchinson, ed., *B-17 Memories: From
     Memphis Belle to Victory* (Bloomington, IN: AuthorHouse,
     2014), 29.
     Rick Atkinson, *The Day of Battle: The War in Sicily and Italy,
     1943–1944*, Liberation Trilogy, vol. 2 (New York: Henry Holt
     and Company, 2007), 13.

On Wings of Eagles

47   *The Eagles*, class of 44-D, yearbook of Eagle Pass Army Air Field.
Cora Montgomery, *Eagle Pass; or, Life on the Border* (New York: George G. Putnam & Co., 1852), 96.
Henry Gannett, *The Origin of Certain Place Names in the United States*, 2nd ed. (Washington, DC: Government Printing Office, 1905), 112.

48   *Aristotle's History of Animals. In Ten Books*, trans. Richard Cresswell, book 9 (London: Henry G. Bohn, 1862), 253.
Patricia Anne Lynch, *Native American Mythology A to Z*, revised by Jeremy Roberts (New York: Chelsea House Publishers, 2010), 36.
"Tito, the Steer," in class book for advanced pilot classes 43-F and 43-G at the Eagle Pass Army Air Field, 33.

49   *The Eagles*, class of 44-E, yearbook of Eagle Pass Army Air Field.

50   "Dedication" in *The Eagles*, class of 44-F, class book of Eagle Pass Army Air Field.

53   The number *630* is documented in Vaughn's monthly chaplain report.
Colonel Raymond K. Bluhm, Jr., ed., *World War II: A Chronology of War* (New York: 2003), 453.

54   Ibid., 455.
Bill Yenne, *Hap Arnold: The General Who Invented the U.S. Air Force* (Washington, DC: Regnery Publications, Inc., 2013), subtitle.
Ibid., introduction.

56   1st Lt. Vaughn H. MacArthur 368th F.A. Chap-Res., "Military Discipline as an Aid to Character Building." From Vaughn MacArthur's chaplain file at the National Archives, St. Louis, Missouri.

57   Eugene Haynes Butler, *Life Story of Eugene Haynes Butler: Fighter Pilot*. From the family records of Eugene Haynes Butler; accessed August 18, 2008, www.familysearch.org/service/records/storage/das-mem/patron/v2/TH-904-65005-119-30/dist.txt?ctx=ArtCtxPublic.

57   "Dedication" in *The Eagles*, class of 44-C, yearbook of
      Eagle Pass Army Air Field.

The Crack-Up
59   *Republic P-47 Thunderbolt Pilot's Flight Operating Manual*
      (Fairfield, OH: U.S. Army Air Force, 1943), 20-26.
61   Brig. Gen. Kennard R. Wiggins Jr., *Images of Aviation: Dover
      Air Force Base* (Charleston, South Carolina: Arcadia
      Publishing, 2011), 9.
62   *Hazardous Waste Ground-Water Task Force: Evaluation of
      Dover Air Force Base, Dover, Delaware* (United States
      Environmental Protection Agency, 1988), 26.
      P-47 Thunderbolt Pilot's Flight Operating Manual (Fairfield,
      OH: The U.S. Army Air Force, 1943), 23.
64   Wiggins, *Images of Aviation: Dover Air Force Base*, 25.

Blitzkrieg
68   Dr. Harold O. Whitnall, quoted in "Mount Vesuvius Eruption
      Goes On; Is Less Intense," *Lawrence (KS) Daily Journal-
      World*, vol. 88, no. 71, March 23, 1944, 3.
      "The Fifteenth Air Force," accessed July 17, 2018,
      www.15thaf.org/index.htm.
69   Tony Bartley, *Smoke Trails in the Sky: The Journals of a
      Battle of Britain Fighter Pilot*, rev. ed. (Cheshire, UK: Crecy
      Publishing Limited, 1997), 9.
      Stephen E. Ambrose, *D-Day, June 6, 1944: The Climactic
      Battle of World War II* (New York, NY: Simon & Schuster,
      1994), 275.
70   "Conditions in Normandy," Memorandum, June 3, 1944
      (Dwight D. Eisenhower's Pre-Presidential Papers, Box 137,
      "Crusade In Europe," accessed July 19, 2018,
      eisenhower.archives.gov/research/online_documents/d_day.
      html), 2.
      "Report of the 8th Air Force, Normandy Invasion, June 2–17,
      1944," Walter Bedell Smith Collection of World War II
      Documents, Box 48, "Eighth Air Force Tactical Operations in
      Support of Allied Landings in Normandy 2nd June–17th June

1944," accessed July 19, 2018, eisenhower.archives.gov/
research/online_documents/d_day.html, 4.

Ambrose, *D-Day, June 6, 1944*, 275.

Samuel Eliot Morison, *The Two-Ocean War: A Short History
of the United States Navy in the Second World War* (Boston:
Little, Brown and Company, 1963), 387.

Major General Bruce Jacobs, AUS (Ret), "World War II:
Stateside and European Theater," in *U.S. Army: A Complete
History*, ed. Raymond K. Bluhm, Jr. (Arlington, VA: The
Army Historical Foundation, 2004), 460.

Ambrose, *D-Day, June 6, 1944*, 275.

72 "Unholy Order of the Piggy-Back," 33rd PRS Online:
Memories, accessed August 3, 2018, www.33rdprs.
photorecon.org/html/memories/piggy.html.

73 Justus D. Doenecke and Mark A. Stoler, *Debating
Franklin D. Roosevelt's Foreign Policies, 1933–1945*
(Lanham, MD: Rowman & Littlefield Publishers, Inc.,
2005), 135.

Martin Caidin, *Fork-Tailed Devil: The P-38* (New York:
iBooks, 2016), n.p.

74 Geoffrey Hayes, *Crerar's Lieutenants: Inventing the
Canadian Junior Army Officer, 1939–45* (Vancouver:
UBC Press, 2017), 177; *The 461st Liberaider*, 24, no. 1,
June 2007, 29.

75 Adolf Hitler, quoted in Mark Riebling, *Church of Spies:
The Pope's Secret War Against Hitler* (New York: Basic
Books, 2015), 177.

Thundering Herd

79 *Alexandria Daily Town Talk*, Alexandria, Louisiana, October
14, 1944, vol. LXII, no. 182, 1.

83 "Fort Knox" in *The Encyclopedia of Louisville*, ed. John E.
Kleber (Lexington, KY: The University Press of Kentucky,
2001), 311.

W. Warren Wagar, *H. G. Wells: Traversing Time* (Middletown,
CT: Wesleyan University Press, 2004), 138.

Leonardo da Vinci, "Letter of Application to Ludovico
Sforza," in eds. Lawrence S. Cunningham, John J. Reich, and

83   Lois Fichner-Rathus, *Culture & Values: A Survey of the
      Humanities, Volume 2*, 9th ed. (Boston: MA: Cengage
      Learning, 2018), 418.
      Leonardo da Vinci, Codex Arundel (London: British
      Museum), Folio 1030, Drawing No. 68.
      Spencer C. Tucker, *Tanks: An Illustrated History of Their
      Impact* (Santa Barbara, CA: ABC-CLIO Books, 2004), 18.
84   Christopher R. Gabel, *United States Army GHQ Maneuvers of
      1941* (Washington D.C.: Center of Military History United
      States Army, 1991), 9.
      David G. Anderson and Steven D. Smith, *Archaeology,
      History, and Predictive Modeling: Research at Fort Polk,
      1972–2002* (Tuscaloosa, AL: The University of Alabama
      Press, 2003), 48.
85   Sharyn Kane and Richard Keeton, *A Soldier's Place in
      History: Fort Polk, Louisiana* (Tallahassee, FL: Southeast
      Archeological Center, National Park Service, 2004), 53.
      Lauren C. Post, ed., Samuel H. Lockett, *Louisiana As It
      Is: A Geological and Topographical Description of the
      State* (Baton Rouge, LA: Louisiana State University
      Press, 1969), 47.
86   H. Paul Jeffers, *Command of Honor: General Lucian
      Truscott's Path to Victory in World War II* (New York,
      NY: New American Library, 2008), 49–50.
      Charles R. Leach, *In Tornado's Wake: A History of the
      8th Armored Division* (Nashville, TN: Battery Press, 1992).

The Fork-Tailed Devil

94   John Stanaway, *P-38 Lightning Aces 1942–43* in the Osprey
      Aircraft of the Aces Series, 120 (Oxford, UK: Osprey
      Publishing, 2014), 7.
99   Kevin A. Mahoney, *Fifteenth Air Force Against the Axis:
      Combat Missions over Europe During World War II* (Lanham,
      MD: Scarecrow Press, Inc., 2013), 249.
      Bea Lewkowicz, *The Jewish Community of Thessaloniki: An
      Exploration of Memory and Identity in a Mediterranean City*
      (doctoral dissertation, London School of Economics, 1999), 2.

See also Antonio J. Muñoz, *The German Secret Field Police in Greece, 1941–1944* (Jefferson, NC: McFarland & Company, Inc., 2018), 98.

Frederick B. Chary, *The Bulgarian Jews and the Final Solution, 1940–1944* (Pittsburgh, PA: University of Pittsburgh Press, 1972), 195. See also Mark Mazower, *Salonica, City of Ghosts: Christians, Muslims, and Jews 1430–1950* (New York: Vintage Books, 2006), 411.

100 Royal C. Gilkey, "'Little Friends': The 49th Fighter Squadron" in *AD-LIB*, newsletter compiled by Bob Karstensen, no. 34 (Marengo, IL: 451st Bomb Group [H] Ltd. Publication, 2001), 16.

101 "Honor Roll 14th Fighter Group Jun 1941–Nov 1945," (rev. August 31, 2008, accessed August 8, 2018, raf-112-squadron.org/14thfghonor_roll42_43.html.

102 The P-38 Lightning nicknames and their pilots are documented at The Modelling News (accessed August 9, 2018), www.themodellingnews.com/2012/12/kageros-december-items-five-of-best.html.

Robert Peczkowski, *Lockheed P-38J-L Lightning,* ed. Roger Wallsgrove (Sandomeriz, Poland: STRATUS s.c., 2013), 24.

103 "The Lockheed P-38 Lightning: Most Versatile Plane of World War II" (Newsbureau, Burbank, Ontario, CA: Lockheed Aircraft Service Company [a division of Lockheed Aircraft Corporation], n.d., from the personal collection of David W. MacArthur), 6.

Mantelli, Brown, Kittel, and Graf, *Lockheed P-38 Lightning, Bell P-39 Airacobra, Curtiss P-40* (Port Camargue, France: Edizioni R.E.I, 2017), 42.

104 Captain Arthur W. Heiden, quoted in Martin Caidin, *Fork-Tailed Devil, the P-38: The Full Story of the Best American Fighter Plane of World War II, and the Men Who Flew It* (New York: Brick Tower Press, iBooks, 2012), n.p.

"The Lockheed P-38 Lightning: Most Versatile Plane of World War II" (Newsbureau, Burbank, Ontario, CA: Lockheed Aircraft Service Company, n.d.), from the personal collection of David W. MacArthur, 4.

Strafing Salonika

111 Second Lieutenant Paul J. Murphy, Missing Air Crew Report 9031, October 9, 1944.

112 Second Lieutenant John R. McCullough, Missing Air Crew Report 9090, October 9, 1944.

114 "Honor Roll 14th Fighter Group Jun 1941–Nov 1945," (rev. August 31, 2008, accessed August 7, 2018, raf-112-squadron.org/14thfghonor_roll42_43.html.
Robert C. Groom's eyewitness statement, Missing Air Crew Report 9046. See also Frank C. Foster, *United States Army, Medals, Badges and Insignias* (Fountain Inn, SC: Medals of America Press, 2011), 133.

"For You, the War Is Over"

118 Missing Air Crew Report 9032, October 7, 1944 (declassified 1973).

Smokey Joe

133 MIRS (London Branch), "Axis Concentration Camps and Detention Centres Reported as Such in Europe," Supreme Headquarters Allied Expeditionary Force, Evaluation and Dissemination Section G-2, Counter Intelligence Sub-Division, December 12, 1944, 97.

141 "Case No. 47. The Hostages Trial: Trial of Wilhelm List and Others. United States Military Tribunal, Nuremberg. 8th July, 1947, to 19th February, 1948" (printed in Great Britain under the authority of His Majesty's Stationery Office by Cole and Co. [Westminster] LTD), 80.

Sabotage

144 The name of "Mrs. Anna Bro" is found on Lloyd Bro's registration card (D.S.S. Form 1, Serial Number 156). The name of Bro's wife, "Mrs. Shirley Mae Bro" is found on the Missing Air Crew Report, 9033 (October 6, 1944).

Red Tails

165 "Today Germany belongs to us/And tomorrow the whole
world" is a loose English translation of the German phrase
*Denn heute gehört uns Deutschland/Und morgen die ganze
Welt.* Hans Braumann first wrote this line for his 1932 song
"Es zittern die morschen Knochen" ("The Frail Bones
Tremble"). Three years later, it became the official marching
anthem of the Reich Labor Force, was chanted by the Hitler
Youth Movement, and was used throughout World War II as
Nazi propaganda. See H. W. Koch, *The Hitler Youth: Origins
and Development 1922-1945* (New York, NY: Cooper Square
Press, 2000), 88.

166 At the start of 1943, the Bulgarian government surrendered its
Jewish residents to the Nazi police. Spearheaded by
Alexander Belev, the commissar of Jewish Affairs in
Bulgaria, a total of 7,300 Macedonian Jews were arrested and
detained without food or sanitation provisions in the large to-
bacco factory in Skopia. With the exception of the 165 Jews
released from the factory, most of the detained Jews were
transported from Skopia to Treblinka, Poland, in late March
1943, where they died in the gas chambers.
"Authorizing the President to Award a Congressional Gold
Medal to the Tuskegee Airmen, H.R. 1259" (Congressional
Record vol. 152, no. 23), H407–H414.

167 "Table depicting number of participants in the Tuskegee
Syphilis Study showing number of patients with syphilis and
number of controlled non-syphilitic patients" (Tuskegee
Syphilis Study Administrative Records, 1929–1972, Record
Group 442: Records of the Centers for Disease Control and
Prevention, 1921–2006, U.S. National Archives and Records
Administration), National Archives Identifier: 281642.

168 Charles F. Francis, *The Tuskegee Airmen: The Men Who
Changed a Nation,* 146–147; Daniel L. Haulman, *Tuskegee
Airmen Chronology,* Organizational History Branch, Air Force
Historical Research Agency, Maxwell AFB, AL 36112-6424,
May 11, 2016; and Missing Air Crew Report 8980.

The Flying Toolshed

173 Conrad Malte-Brun, *A System of Universal Geography; or, a Description of All the Parts of the World on a New Plan, According to the Great Natural Divisions of the Globe; Accompanied with Analytical, Synoptical, and Elementary Tables, Vol. II* (Boston, MA: printed and published by Samuel Walker, 1834), 447.

177 Dave MacArthur reported in several interviews that, following his release as a POW, his superiors concluded that he was held for three to five days in Dachau while in transit from Greece to Stalag Luft III. "Course, I didn't know I was even in Dachau until after the war. Everything sort of got pieced together during debriefing," he said. But, since his time in Dachau wasn't able to be confirmed, Dave described the conclusion as "purely conjecture" and, over the years, rarely mentioned Dachau by name. Though it contains inaccuracies regarding Dave's wartime experiences, the article "Dachau to Duke: An Amazing Saga of One of Your Classmates Who Saw the Inside of Dreaded Horror Camp—Dachau" was published by Pete Maas in the Duke University *Duke and Duchess* magazine several years after Dave's release. However, it is also possible that the cell in which Dave breathed the poisonous gas was in any number of other concentration camps.

"Lieutenant David W. MacArthur, 0-714466"

182 "American Prisoners of War in Germany," prepared by Military Intelligence Service, War Department, 1 November 1945, Dulag Luft.

184 Part II: Capture—Article 5, in "Convention Relative to the Treatment of Prisoners of War, Geneva, 27 July 1929," accessed September 9, 2018, ihl-databases.icrc.org/applic/ihl/ihl.nsf/ART/305 430006?OpenDocument.

185 Raymond F. Toliver, *The Interrogator: The Story of Hanns Scharff, Luftwaffe's Master Interrogator* (Fallbrook, CA: Aero, 1978), 138.

186 Gerhardt B. Thamm, "Documentation Exploitation After WWII: The Potsdam Archive: Sorting Through 19 Linear

Miles of German Records," in Andres Vaart, ed., *Studies in Intelligence, Journal of the American Intelligence Profession, Unclassified articles from Studies in Intelligence,* Vol. 58, no. 1 (Washington D.C.: Center for the Study of Intelligence, Central Intelligence Agency, March 2014), 5.

The Great Escape

191 "American Prisoners of War in Germany: Stalag Luft 3, Air Force Officers," prepared by the Military Intelligence Service War Department, November 1, 1945.

198 J. Ted Hartman, *Tank Driver: With the 11th Armored from the Battle of the Bulge to VE Day* (Bloomington, IN: Indiana University Press, 2003), 39.

The Continent

203 Charles R. Leach, *In Tornado's Wake: A History of the 8th Armored Division* (Nashville, TN: Battery Press, 1992).

The Search

229 For more information on the roles and experiences of WWII chaplains, see Lyle W. Dorsett, *Serving God and Country: United States Military Chaplains in World War II* (New York: Dutton Caliber, 2013).

This account is from the personal recollections of Vernon Miller, Battalion Supply Clerk of Headquarters, 130th Ordnance Battalion, Eighth Armored Division, in e-mailed correspondence on January 30, 2018.

231 This account is from "A Christmas Story—Told by Salomon Braus," c. 1997, from the personal records of David W. MacArthur. "Salomon Braus" is a pseudonym used to protect the involved parties.

233 "Nazi Camps," *Holocaust Encyclopedia*, United States Holocaust Memorial Museum, encyclopedia.ushmm.org/content/en/article/nazi-camps.

Stars and Stripes

234 "History of *Stars and Stripes* Newspaper," starsandstripes.
newspaperarchive.com/history.
From information received via e-mailed correspondence with
Catharine Giordano, Supervisory Archivist & Licensing and
Permissions Representative, Stars and Stripes—Library &
Archives, on September 28, 2018, and from "The Stars and
Stripes—European Edition,"
www.usarmygermany.com/Units/StarsandStripes/USAREUR
StarsandStripes.htm.

Meatloaf and Apple Pie

248 "Follows Hunch, Finds Son," *Armored News*, June 4, 1945.
From Vaughn MacArthur's chaplain file at the National
Archives in St. Louis, Missouri.

Home

254 "Memorial Held for 'Good Joe,' 8th AD Chaplain," *Armored
News*, Monday, July 30, 1945. From Vaughn MacArthur's
chaplain file at the National Archives in St. Louis, Missouri.
Charles R. Leach, *In Tornado's Wake: A History of the 8th
Armored Division* (Nashville, TN: Battery Press, 1992),
chapter 9.

255 "Lorraine American Cemetery," American Battle Monuments
Commission, www.abmc.gov/cemeteries-memorials/europe/
lorraine-american-cemetery#.W4bsXS3MyRs.

Bugout

259 Joe Quinn, "Shattering Air Bombardment Saved 'Trapped'
ROK Unit, U.S. Advisors," n.d. Clipping from the personal
collection of David W. MacArthur.

265 Alexander L. George, *The Chinese Communist Army in
Action: The Korean War and Its Aftermath* (New York:
Columbia University Press, 1967), 9.
In 1953, the Pentagon claimed that 54,260 American lives
had been lost in the Korean War. In the 1994 publication of
"Service and Casualties in Major Wars and Conflicts," the

revised death toll fell to 36,914 and included both "battle deaths" that occurred in Korea (33,652) and "other deaths" (3,262) that included accidents, illnesses, and non-battle fatalities that occurred at the same time but outside Korea. The most recent death toll, according to the U.S. Department of Defense and Veterans Affairs, is estimated at 33,739 "battle deaths," 2,835 "other deaths" (in Korea), and 17,672 "other deaths" (outside Korea).

266 Xiaobing Li, *China's Battle for Korea: The 1951 Spring Offensive* (Bloomingdale: Indiana University Press, 2014), 124.

"U.S. Air Force Recipients of the Army Distinguished Service Cross, All Conflicts," valor.defense.gov/Portals/24/Documents/ServiceCross/Army DSC-AirForce-AllConflicts.pdf.

"David Warren MacArthur," The Hall of Valor Project, valor.militarytimes.com/hero/6879.

ACKNOWLEDGMENTS

In 1946, from his college dorm typewriter, Dave MacArthur set out to record his World War II experiences, noting that "to relate it all in detail would fill a small book, so I am going to cover the important facts."

Filling in the gaps between those facts, however, was possible only with the help of the many people below.

In July 2017, a mere six months after her husband's death, Sharon Kinne MacArthur welcomed me into her home to pore over Dave's extensive collection of letters, newspaper clippings, scrapbooks, interviews, government documents, photographs, flight logs, and other records. Over that week in Florida, and through the countless hours of telephone, text, and Face-Time conversations that followed, she shared delightful anecdotes and poignant memories of her husband's life, and put me in contact with others who did the same. It is because of Sharon's joyful and tireless stewarding of Dave's story that *Lightning Sky* exists.

Special thanks is also due to Captain John J. Shenton, Air Force Reserve and retired American Airlines pilot, who now flies World War II–era planes. Not only did he share personal remembrances of Dave, which added significantly to this story, but with an eagle eye he combed through the manuscript to ensure that my aircraft descriptions and flight sequences ring true. Any errors or oversights are mine alone.

Two records in Dave's personal collection offered particularly valuable insight: John Quinn's 1996 recorded audio interview, and Gwen Merrick's 2008 profile of Dave, which became the basis of her paper entitled "P-38s and Patriotism: A POW's Perspective."

Thanks, as well, to Peter Egeli for sharing his experiences with Dave; to Sharon, Bruce, Bayne, and Denise for reading early versions of the manuscript and offering constructive feedback; and to John B. Thomas III whose father, John Barney Thomas, named and piloted the P-38 Lightning (*Joanne*) in which Dave was shot down on October 6, 1944.

When the manuscript was early in its infancy, CW2 Christopher S. Young, U.S. Army, provided military-related information and useful guidance in contacting members of several different military associations, including the Eighth Armored Division Association, three members of which were especially helpful in gathering information on Vaughn MacArthur: David J. Clare, association historian, whose father, Sergeant Paul W. Clare, served in Company C of the division's Fifty-eighth Armored Infantry Battalion; Vernon Miller, a ninety-five-year-old veteran who served with Chaplain MacArthur in World War II (Headquarters of the 130th Armored Ordnance Maintenance Battalion); and Andy Waskie, Jr., president of the Eighth Armored Division Association, whose father, First Lieutenant Andy Waskie, served as a senior medical officer (also Headquarters of the 130th Armored Ordnance Maintenance Battalion).

Much of my understanding of Vaughn's service during the war is credited to Lyle Dorsett, who graciously directed me to the chaplain files held in Record Group 247 of the National Archives in St. Louis, Missouri, and whose excellent book *Serving God and Country: United States Military Chaplains in World War II* exposed me to the triumphs and challenges many chaplains faced.

The staffs of several entities were especially helpful in securing records, tracking down obscure bits of information, and providing access to material not otherwise available: the staff of the National Archives in St. Louis, Missouri; Catharine Giordano, supervisory archivist of the *Stars and Stripes* Library and Archives; Valeria Pendenza of *L'Osservatore Romano* Archives; and Marco Grilli, secretary of the Prefecture of the Vatican Secret Archives.

Special gratitude is reserved for my agent, Greg Johnson, president of WordServe Literary and longtime champion of the forgotten stories of World War II.

The extraordinary publishing team at Kensington has gone above and beyond in producing a book worthy of Dave's story. I am especially indebted to president and CEO Steven Zacharius; vice president–general manager Adam Zacharius; vice president–publisher Lynn Cully; director of communications Vida Engstrand; copy chief Tracy Marx; assistant editor Norma Perez; copyeditor Brenda Horrigan; editor in chief of Citadel Michaela Hamilton, who arranged the first contact with Sharon MacArthur; creative director/trade Kristine Noble, who designed *Lightning Sky's* beautiful cover; production director Joyce Kaplan; production editor Arthur Maisel; and my publicist, senior communications manager Ann Pryor.

I owe deepest thanks to Wendy McCurdy, my inestimable editor and a master of story craft, who first saw value in telling Dave's story and who shepherded this project with skill, finesse, and patience edging on the infinite.

Only days after Dave's death, Wendy's brother, Jeffrey McCurdy, sent a photo of Dave's obituary to his sister with the words, "Somebody needs to write this story." For allowing me to be that "somebody," and for entrusting me with so precious a narrative, I remain enormously grateful.